Destination Eden

Fruitarianism Explained

Mango Wodzak

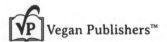
Vegan Publishers™

Published by:
Vegan Publishers
Danvers, MA
www.veganpublishers.com

Cover and text design by Jonathan Byrd

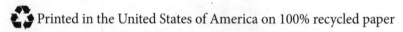 Printed in the United States of America on 100% recycled paper

ISBN: 978-1-940184-25-8

Table of Contents

Acknowledgements

My gratitude goes out to my loving partner, Květa, for her feedback and patience with me as I struggled to finish this book, forever tweaking and fine-tuning it. Never fully content. I finally realized I likely never will be.

My thanks also go to Zalan, Konrad, and Esther, for their help in proofreading, and initial feedbacks of draft versions.

Finally, thanks to everyone else I've met on my journey, who have offered insight and encouragement, and all the great people out there who are ready for the message and vision I'm about to share with you.

Dedication

This book is dedicated to the wave of awakening consciousness, to those who can see past society indoctrination, and who, looking positively forward to a love and light-filled world, are prepared to make the lifestyle changes necessary for such a vision.

Disclaimer

Let it be clear from the onset, that I am neither a doctor nor a qualified nutritional expert. I left school at the tender age of sixteen and am happy to have had no other formal education. Whether or not you choose to share or act on the thoughts and visions I present throughout this book is your personal matter.

I have no desire to take on any obligations for your resulting personal choices. If you are in need of someone to take responsibility for your decisions in life, then I suggest you consult a suitable, accredited professional.

My advice within this book is that you follow or do what you truly believe to be correct. That you read my words with an open mind, give them time to digest, and then reach your own conclusions.

Naturally, I cannot deny, I would like for readers to share my vision, but my desires should not interfere with your own judgments. I'm not here to tell you how you yourself should or should not live your life, what you should or should not believe, or how you should or should not eat. These are your choices, and you alone must decide what you feel is right for you.

The purpose of this book is for me to explain why I have made the choices I have.

Make of it what you will.

Preface

By Kveta

Lies have corrupted moral values, and through those lies Eden has been lost. Eden has been stolen from us, and our lives stripped of love. To this day, Eden is a fantasy instead of a reality. For most of us Eden does not exist. Puzzled minds delivering confusion and creating pain. Eden is deserted. This is the essence of this book. Mango pinpoints the reasons why.

However, the essence of this book is also joyous. Mango shares his own experiences and a radiant future vision. He hopes to inspire others, and explains how to free all existence from sorrow, embracing instead a happy, love filled life.

- Květoslava Martincova - July 2013

By Zalan

Mango's love and life shine brightly throughout this book of sunlight. Refreshing and powerful, a sacred recipe for love, peace, goodwill, and sustainable living. A deeply private, honest, and simple account of the journey of his inner child, confronted by violence and suffering, seeking explanations, and modifying his life choices and actions in response.

From an early childhood honed by Monty Python and The Goodies, as a teenage altar boy flirting with reality, science fiction, and a creative lifestyle, to an adult struggling to combine ethics and compassion within his social and cultural setting, we now witness the fully developed, integrated, holistic, spiritual nature of a self-professed Eden Fruitarian!

Couched in the genre of 'isms,' the novel text is compiled from well-proven experience, with ample portions of reality, research, and imagination from someone with his feet on the ground. A generous lifetime of home-grown wisdom, some colorful language, a dash of excitement, a couple licks of irony, garnished with revelations, and a good sprinkle of fun and humor. Definitely one of the most easily digestible books I have read.

Shoes off, step inside, leave your prejudice at the door, settle down in a comfortable chair with a bowl of Mango's fruit salad, and hold on tightly to your essentials, while you join Mango on his flight to fruity freedom; be prepared to gorge the lot.

Life as it is, naked and exposed, our social and cultural norms questioned, while nature's secrets are disclosed; a new benchmark in un-fired un-cooking, with an abundance of ideas that will fire your brain, and a choice of words that will melt in your mouth, this unique tome is serious fun, and funnily serious.

It is an honor to introduce this work, which I consider a major contribution from a brave heart and soul, with the capacity to influence each of us willing to walk the path of personal responsibility while traversing the puddles and slippery slopes of self-awareness. It is very likely your life is not here right now. If you are interested in free expression, personal ethics, philosophy, animal rights, health, and the challenge of culinary reading, see how you go digesting it slowly, but be prepared to be stimulated. My fingers are still burning from turning the pages.

In documenting his quest for light, Mango has produced a brilliant tale. As self-confessed fruit-nuts, he and his partner,

Květa, are without peer. Part of their recent journey to Eden may be viewed at "Pure Fruit"[1] on Vimeo. Welcome to Eden Fruitarianism, a seamless trajectory to paradise, and many happy returns!

- Zalan Glen - August 2013

Introduction

Hi, my name is Mango, and I'm a fruitarian.

In claiming to be fruitarian, I don't mean one of the multitude of fruitarians out there who are scared to eat only fruit and advise people that if they do so they will become deficient, get sick, and ultimately die an early, drawn-out, demineralized death. I'm not one of those fruitarians who advise people to make fruit an arbitrary percentage of their diet and then the rest, greens, nuts and seeds. Nor am I one of those fruitarians obsessed with calorie counting, and analyzing the handful of separately classified scientific elements purportedly microscopically available within each fruit.

I am also not one of the vast bulk of fruitarians who sully fruits by referring to them degradingly as sugars and fats, nor one whose focus is predominantly on fitness and one's ability and prowess in triathlons or iron-man competitions. Although there are plenty of those kinds of fruitarians out there, I'm not one of them!

I'm currently one of a very rare minority of fruitarians who advocate one hundred percent raw, fresh, ripe, fruit, fruitarianism. I differentiate between nuts and fruits, dried, shriveled fruit and fresh, ripe fruit. I don't believe greens should ultimately form part of one's diet, and I don't drink sugar cane and/or coconut juice.

I believe in practicing what I preach. I talk of what I believe to be true fruitarianism as it is meant to be, and the understanding of that also happens to be the key toward (re-)establishing Eden on this planet.

I openly admit that parts of this book are about beliefs, and you will not fail to notice that I actively shun scientific evidence to support these beliefs. So if it happens that you've stumbled upon this book and you're interested in the concept of fruitarianism but are expecting to find pages of nutritional tables coupled with the latest scientific viewpoints to support fruitarianism, then I should come right out and tell you now that these expectations will definitely not be fulfilled, at least not through this book.

I will not be breaking down foods into their supposed constituent microscopic elements, nor will I be balancing pHs, complimenting [sic] proteins, tracing elements, food combining, or listing recipes. I will however, explain why I actively and consciously eschew such an approach to fruitarianism.

One of many things that appear clear to me is that the raw food movement has been both commandeered and corrupted by a business-minded minority who quote nutritional science to suit their needs and make it seem as though eating raw is overly complex. By employing a certain amount of tactical scaremongering, they hoodwink the uncomprehending gullible into believing that they must first study hard under the greased palms and guidance of an expert, or attend countless costly seminars. The alternative being to surely succumb to a variety of cleverly christened deficiencies.

I've read enough books to know that somehow or other, nutritional science can be used to prove many conflicting theories. I believe this is because such research is far from complete, and is therefore open to being interpreted in more than one way. Consequently, nutritional science can be easily manipulated and used to convince people of whatever the author's desire. Personally, I find the idea of following a nutritional science, based on a largely incomplete set of facts, to be rather absurd, and sadly often so complexly absurd that it borders on the dangerous. Instead, I believe that when it comes to food, science is a red herring and our one true guideline should be ethics. My belief is that love is the one true force; it alone is responsible in keeping the universe ticking. Once this concept has been fully embraced, the innate intelligence of the body, with its own inner physiological wisdom can alone be fully trusted to guide us toward our true individual nutritional needs at any given moment in time.

Living solely on a fruit diet, following the guidelines of Eden Fruitarianism, does not involve enormous amounts of prior knowledge and long laborious PhD dissertations and discourse. This book explains why I believe that to be true.

In case you are at a stage where the supposed facts, figures, and statistics of a purportedly proven science seem essential, or should you have already decided that my approach is too ethereal for your own personal liking, then there are plenty of other books for you to choose from. If you would rather hear the opinion of someone who has been indoctrinated and/or has a string of letters after their name, then I suggest you forget this book and look elsewhere. Exercise free will and pass this book onto someone who you feel could appreciate it more, or else burn the book for fear that the gullible may be corrupted.

In light of the overall absence of nutritional pseudo-science, I am, however, no less confident in my stance, and by no means suggesting to be taken any less seriously. I merely state this as further means of a disclaimer. Again, whether you agree with my musings is for you alone to decide. I have made my choices and am happy with them; I am even open to changing them should I ever be compelled enough to do so.

I write both from plenty of experience, and from the overall success of my persistence. Do not expect though, for this book to be solely full of ramblings on food. Rather, it is a collection of thoughts, many of which happen to be, by their very nature, food-related.

When this book first began taking form back in the year 2000, I tentatively christened it, "Destination Eden - Barefoot Fruitarianism Explained." At the time, I believed that the addition of the word "barefoot" was sufficiently descriptive to distinguish this particular brand of fruitarianism from the evermore-popular "I fear fruit" fruitarianism. But over the years, since I began writing, I've noticed that even among the other self-christened fruitarians, there are also those who call themselves barefoot fruitarians. Yet they regularly don flip-flops or Vibram FiveFingers shoes, and thus add further confusion to the already chaotic mayhem of non-descriptive titles. Clearly I needed to come up with a better distinguishing word, one that could intelligibly separate the wheat from the chaff. Thus the phrase "Eden Fruitarianism" was born, and here's to hoping that it too will not fall victim to semantic abuse.

There are times however, where I will still refer to the philosophy in this book as barefoot fruitarianism, or just plain fruitarianism, and at all times will be considering the titles synonymous with the freshly minted Eden Fruitarianism.

The philosophy and lifestyle of Eden Fruitarianism is multi-tiered, and although ultimately not too difficult to grasp, it does require a full understanding of each of its various levels in order to succeed. The purpose of this book is to shed light on each of those levels, with the hope that in doing so, you too may choose to embrace the age-old philosophy of fruitarianism.

*Oh, Great Spirit,
whose voice I hear in the winds
and whose breath gives life to all the world, hear me.
I am small and weak.
I need your strength and wisdom.
Let me walk in beauty and make my eyes
ever behold the red and purple sunset.
Make my hands respect the things you have made
and my ears sharp to hear your voice.
Make me wise so that I may understand
the things you have taught my people.
Let me learn the lessons you have hidden
in every leaf and rock.
I seek strength, not to be superior to my brother,
but to fight my greatest enemy - myself.
Make me always ready to come to you
with clean hands and straight eyes,
so when life fades, as the fading sunset,
my spirit will come to you
without shame.*

American Indian - Lakota - Chief Yellow Lark - 1887

Section One

THE NEED FOR CHANGE OF CONSUMER HABITS

Our quest, our earth walk, is to look within, to know who we are, to see that we are connected to all things, that there is no separation, only in the mind.

- Lakota Seer

Chapter 1
Omnivorism

Slippery Slope

I've heard it said by some that marijuana is the first step taken on the road to heavy drugs. It is my belief, however, that the first step is taken long before such an eventuality. Long before we taste our first drops of alcohol, or experiment secretively with our first wheeze on a cigarette, we are on that road. It's weathered even before we suffer our first coffee, or embark upon the fried crispy boulevard of denatured junk food. The path we are generally set upon from birth leads pretty much inevitably there, sooner or later. The only thing is, we often reach there without ever acknowledging the fact.

If it isn't clear already, I'm taking liberties with the definition of heavy drugs, and referring also to the synthetic, chemical-based, mind-altering drugs gleefully dispensed by the pharmaceutical industry.

I think it indubitably correct that, as the old adage goes, humans are creatures of habit. Indeed, habits can become so ingrained that from our own individual perspectives, we often become simply obliviously unrecognizing of their presence.

This is surely the case with certain lifestyle choices that are unwittingly thrust upon us at an early age. Often so young that they, more often than not, remain unnoticed, unquestioned, and unchallenged throughout the duration of one's entire earthly existence.

Let me give an example. We all enter this world as potential barefoot fruitarians. Fruit has that upfront mixture of appeals, vibrant colors, the tantalizing textures, and deliciously sweet flavors that, provided corruption and addiction are not already imbued, are instinctively appealing to all children. Alas, it's a path few of us are privileged or permitted to have at such an early age. Instead, it rarely takes too long following nascence, before we find ourselves involuntary shod omnivores. Rarely is such a state of affairs ever cross-examined.

Shod Omnivore

"Shod omnivore," you say? Yes, this, I propose, is the antipode of the barefoot fruitarian—the path a vast majority of us are placed upon, and the one the vast majority of us follow until our untimely graves. Rarely also is such a path ever contested, and yet rarer is it forsaken.

Shod Omnivorism pretty much sums up the current state of the world. I find the word 'shod' deeply descriptive; it conjures such an archaic, barbaric feeling of being shackled and tamed, captive and downtrodden. Omnivorism is the concept that every living and non-living thing on this planet is a potential meal, and God forbid that anyone should question another's right to eat whatever, whomever[2], and however they desire. Shod and Omnivorism together truly emphasize the sad disequilibrium of the dominant prevailing human mindset.

One thing that consumed drugs all have in common is our initial, natural aversion toward them. The first mouthful of alcohol we drink is generally followed by an involuntary grimace. The first puff on a chemical-laden cigarette is often followed by a cough and splutter as the body tries to repel the alien pollution thrust upon it. Our first coffee and tea are generally also greeted somewhat similarly. Of course, it is frequently the case that despite these initial reactions, we push on past them until addiction is formed. Cooked food, although noticeably less recognized as addictive, evokes no less an initial reaction. Think of all those babies whose faces screw up in displeasure, vainly attempting rejection of the denatured slop thrust upon them, and their hours spent crying from stomach pains. By the time they are advanced enough to linguistically voice their lack of desire for such foods, they are, alas, already well hooked.

Unlike other omnivores that basically eat their food as nature provides, the shod omnivore looks at anything as being a potential meal. We start off our days with cooked cornflakes covered in cooked (pasteurized) milk, or greasy fried bread with birds' eggs and thinly sliced pig belly flesh. For lunch, it's pulverized baked wheat with cooked pasty spreads smeared on top, and for dinner, fired animal fleshes with potatoes. Not only do we make a habit of all this, but we go about telling everyone how healthy it is too!

Thus, it is highly likely that this sad disequilibrium of which I speak will be altogether lost and unclear to many, like spraying perfume on a pig; hence, here I sit now, attempting to clarify the reasoning behind my conclusions. Hopefully after reading through this book, light may be shed, and who knows, the proverbial truffula seed (made illustrious through the words of Dr. Seuss in his book The Lorax,[3] my all-time favorite children's book) may be sown.

I am an optimist, and at least some of the seeds disseminated will surely fall on fertile soil, take root, and sprout.

Read on!

Chapter 2
Vegetarianism

At some point in the past, I temporarily hypothesized that the reason why the world was not predominantly vegetarian was essentially a question of intelligence. Indeed, a two-decade study that began back in the 1970s, which had over 8,000 participants, seemed to indicate a correlation between higher intelligence and vegetarianism.[4] However, not knowing all details, or factors present, I think it a little unwise to jump immediately to conclusions.

On life's path, I have encountered numerous flesh-eaters of strikingly high intelligence, who have apparently never stopped to consider the consequences of their omnivorism. This initially led me to believe that the missing factor was more one of ignorance than unintelligence.

It seemed a logical assumption, but with time it became pretty obvious that, even when faced with the blatant facts, most people, regardless of their IQ level, are still unable to see any need to rethink their consumer habits.

This made me delve further, to reconsider possible reasons for this apparent failing of humanity. Clearly there must be another missing ingredient, something else that humans are lacking. I believe that missing ingredient to be empathy.

Empathy

There comes a time when everyone should seriously empathize.

Wikipedia defines empathy as "the capacity to recognize feelings that are being experienced by another sentient or semi-sentient being." Empathy is a prerequisite for experiencing compassion, and compassion is precisely what this world is most in need of. It's the crucial emotion required to help free the world from the thralls of depravity[5] in which it finds itself ensnared.[6]

Actually, in retrospect, it's a good thing that compassion relies on empathy rather than intelligence. We are all capable of compassion, regardless of where we intellectually stand! Intellectually, many people know roughly what goes on behind closed slaughterhouse doors, but even though there is a vague surface knowledge, they really don't see what is happening on a deeper level.

In general, we have a very strange relationship with species other than our own. We may selectively nurture feelings and form relationships with particular animals, fully acknowledging their uniqueness in the process. Bonds are often formed that make it unthinkable for someone to harm their chosen animal friend. Cats and dogs are prime examples of animals that humans regularly become attached to. They are taken in as members of the family and, to a lesser (or greater!) extent, are treated as such.

Being an alien observer, one might think that forming such attachments would help humans to have a better understanding of the uniqueness of each and every individual animal, regardless of species. Yet for some reason, the majority of us are seemingly unable to think outside the narrow box we were placed into. We may see the distinctiveness of a dog's personality, and appreciate that the dog is in many ways similar to us—in need of companionship, affection, and other general life comforts and necessities. We may see and relate to the sadness of the dog when it is unfairly treated, abused, and/or malnourished. Indeed, we have even established organizations that have made it unlawful to treat such animals unjustly.

In moving away from the few chosen species, an altogether different story is revealed. Take the pig, for example, which in many ways is quite similar to a dog. In fact, by some accounts, they are said to be of higher intelligence than dogs. The minority of people who have nurtured a relationship with one can testify as to how they can be equally loyal and faithful as any dog. They can be just as playful, just as humorous, just as protective, just as loving. Like us and dogs, pigs will do their utmost to avoid both intentional and unintentional harm to themselves, as well as others they have formed attachments to. They will cry when cold, fearful, lonely, or suffering. And yet, generally the only place they receive in our hearts is one of added artery fat. In contrast to appearing wide-eyed at the breakfast table, waiting to be petted and treated, they appear on the breakfast table, with their eyes removed and their bodies bloodily butchered into rashes.

In order to satisfy the appetite of the masses, the life of a western pig is filled with far more suffering and cruelty than the ASPCA or RSPCA would ever permit if the victim were a cat or dog. This is a harsh truth that the majority of us are unwilling to face and the reason why time needs to be set aside to practice empathy.

Just why there is this blatant discrimination against pigs and other animals bred for consumption is altogether unclear. On the one hand, we attempt to instill a sense of morality and justice into our children in regards to other animal species—that we should treat them with dignity and respect. On the other hand, there is this generally accepted attitude of "animals are ours to do with as we please."

By way of demonstration, children may be taken to local parks where they are good-naturedly, yet ignorantly, taught to feed the ducks stale white bread, while generally being told to keep their distance and respect their space. Meanwhile, back at home we are taught how to fry the wings of chickens (and ducks!), and even how cows' bottoms should best be tenderized and seasoned, while the ribs of a pig, with a violently bloody history, sit fully disrespected and festering in the microwave. This is a sort of moral schizophrenia known in psychology as:

Cognitive Dissonance

Cognitive dissonance,[7] discomfort experienced when simultaneously holding two or more opposing beliefs, is the reason why meat eaters get so embarrassed, anxious, flustered, and often irrational when these conflicting values are pointed out to them.

Make it a cat's bottom, the ribs of a dog and the pale tender flesh of young puppies, and people will become furious! I've seen cases on the Internet of a dog being locked in a car for several hours, and comments from people are so enraged that they threaten physical violence toward the perpetrator, some even crying out for the owners to be faced with the death penalty! These same people will likely not think twice about the animals on their plates. In fact, they usually react more strongly toward such events (dogs locked in cars) than do most vegans. The often unconsidered, deep-seated reasoning behind their indignation is to draw away attention from their own wrongdoings toward animals, and focus it elsewhere. If they shout loud enough, then perhaps people won't notice that they themselves are behind far worse animal treatment!

Actually, just recently I stumbled upon an online newspaper article[8] about an abattoir worker who, in claiming not to be able to afford veterinary costs, cut the throat of his own puppy. The result?

Five months in prison! And to think, with society's blessing, he'd made his living from slaughtering!

Many of us have had the experience of admiring a field full of sheep and newborn lambs. They are so undeniably cute and full of fluffiness, with such an apparent, eager zest for life. We point them out to our children, joyfully observing their frolicsome glee, and then go home to eat lamb.

Chopped!

And yet, rarely, barely, is any of this ever even given a second thought.

Some pranksters in Brazil set up a mock stall in a supermarket, selling fresh sausages. They offered free, taste-testing samples, and then had a machine that would churn out freshly made sausages to sell. As the machine emptied, they would pick up a live baby pig, place it in the machine, and continue churning out fresh pork sausage. Of course, the piglets weren't really being transformed into sausage on the fly, but customers were ignorant of this. Consumer reaction was one of typical cognitive dissonance. Confusion, shock, horror, disgust, and even violence toward the butcher. Watch it for yourself here.[9] Although the video was made to get laughs, it shows the inherent state of mind of the general public, and proves that the lack of ability to understand the true issue at heart is not one of intelligence nor knowledge, but one of denial and refusal to empathize.

I think the Indian-born British comedian, playwright, and vegetarian Spike Milligan summed it up pretty well when, toying with a poem by William Blake, he penned these four short lines of verse:

> *If a robin redbreast in a cage*
> *Puts all heaven in a rage*
> *How feels heaven when*
> *Dies the millionth battery hen?*

Confronted with these incontestably paradoxical attitudes toward our animal brethren, people come up with all manner of illogical reasons to justify human behavior. For instance, some might attempt to justify things by explaining that animals, unlike humans, are basically instinctual creatures. Thus, if we find ourselves empathizing unduly with them, we are guilty of anthropomorphizing—giving them qualities that are clearly unique to us intellectually superior humans.

The truth is, there is naught wrong with anthropomorphizing. Indeed, other species have much more in common with humans than we regularly care to admit. Anthropomorphizing rightly leads to developing more empathy toward members of other species.

Every pet owner can vouch for this. They know that each animal is endowed with its own peculiarities. They know that when a pet looks sad, bored, or hurt it is probably indeed sad, bored, or hurt. There is no rocket science in this. These are far from being uniquely human attributes, and are shared with pretty much every animal out there, regardless of species. One no more has the right to deny such emotions in other species than a solipsist has the right to deny the emotions of other humans. The fact that animals cannot say, "I am hurt," (in any human language) does not mean they do not hurt.

The dark-aged, primitive society we live in loves to push the theory that we need to eat flesh. As much as the industry that produces it may perhaps be a little gory for some, many choose to view it as a necessary and unavoidable fact of life. In seriously challenging this supposed fact, those who propose it are at a loss to adequately defend their stance. Nutritionally, their argument mostly revolves loosely around protein and vitamin B12.

As stated in the Introduction, my preference is not to get involved with a nutritional science that I personally find untrustworthy, but, rather, I intend to stick with simplicity and ethics. A certain percentage of humans in this world live their lives never really partaking in consuming flesh. Estimates vary as to what percentage of the world is vegetarian, but there seems to be a generally accepted agreement that the figure lies somewhere between five and ten percent of the world's population. Their life expectancy, by some estimates, is reportedly higher than that of the average omnivore.[10] Personally, I have no need to seek further evidence. I understand that people can live at least lives of average lengths (or longer), even more healthily, without ever adopting zombie-like flesh eating habits. Therefore, I have no issue in dismissing any claims that one must eat flesh.

In the Australian Dietary Guidelines released in 2013, Australia's top health experts agreed with leading health advisory boards in the US and Canada that well-planned vegan diets are a safe, healthy, and viable option for all age groups.[11]

Another commonly voiced supposed reason as to why it's acceptable to eat animals is that they blatantly exist to be eaten, as this is just nature's way. Bears eat fish, lions eat zebras; look around you! The world is a killing machine. There exists an undeniable food chain, and we are at the top of it. Thus, we have the God-given right to eat anything beneath us.

This argument supposes that because animals eat other animals, we too should be granted the same privileges. Logically, this implies that being indifferent to human suffering should also be an acceptable behavior; a crocodile wouldn't help a drowning human, so why should we? Other animals do lots of things we might consider immoral. There is no jungle law prohibiting any behavior, regardless of how gross we deem it to be. Murder, rape, theft, incest, etc. all happen. So should we argue that everything that happens in nature be allowed within human society? While we are looking up to animals as model citizens to mimic, should we eat our own feces too, like dogs sometimes do? And while, obviously, many animals are natural born carnivores and omnivores, comparative anatomy indicates that we are physiologically herbivore/frugivore.[12]

I say forget the folly of looking up to other species as examples as to how we should or shouldn't behave. Instead, we need to begin to think for ourselves, and there is no more meaningful philosophy than to treat others as we ourselves would ultimately prefer to be treated.

Ask yourself, would it be okay for someone to imprison you, fatten you up by feeding you crap, and then ultimately slaughter and butcher you for supper? Of course not! I would not be happy with this and neither would you, nor anyone else. Now ask yourself why you might find it tolerable for other individuals of other species to endure such treatment and untimely death. Seriously ask yourself, because without empathy, there can be no compassion! If you answer that animals do not share the awareness and intelligence of humans, then surely by this logic small children, the mentally retarded, and comatose people also do not deserve rights to protection, since they also can neither reason nor formulate ethical positions as mentally healthy adult humans are able.

Don't be mistaken and believe I am talking now purely of death and dying. This inevitable fate awaits us all. I am talking strictly about immoral imprisonment of innocent sentience[13] and their final cold-blooded murders.

Speciesism

I am not the first to propose this, but the only genuine reason I can see why such behavior toward other species is sanctioned, is that the bulk of humankind are speciesists. Similar to the chauvinists, whose belief is that the opposite gender is basically inferior, the ageists who discriminate against others because of their age, and the racists whose belief is that certain races or those of certain skin tones are inherently inferior, the speciesists believe it is perfectly normal and just to abuse other beings purely because they are not human. There is, of course, no logic, no rational reasoning, behind drawing such conclusions. There is no logic in deciding what's acceptable and permissible in disregarding the needs of certain selected animal species.

In practice, within the western world at least, there are really only a few predominant species that are singled out and subjected to lives of torment; cows, pigs, sheep, chicken, and fish probably receive the bulk of the abuse. If factory farming for meat of cats, dogs, squirrels, swans, and guinea pigs[14] began in Western Europe, you can be sure that some of the bacon and sausage-gorging public would be out protesting. Although other cultures regularly eat some or all of these animals, everybody draws the line somewhere. Most would balk at the idea of eating dolphin, gorilla, orangutan, or human flesh, but really the differences between the species are minimal, and whether we are a rabbit, horse, chimpanzee, or human, we all have an innate desire to live our lives freely and avoid violation.

I say put aside backgrounds, traditions, and cultures, because they are often unexplored reasons behind the discriminatory choices made. Instead, empathize with your victim, and put yourself in her/his/its place. Would you be content if nobody did anything to deliver you from such a life of evil and torment?

Faced with the Facts

By being forced to truly confront the cold hard facts behind meat pro-duction,[15] I strongly believe that the majority of people, especially in the western world, would cringe and feel disgust and likely also shame.

I think this world would stand to benefit from sending everyone on a slaughterhouse visit at least once in their lives. Or at least if we could all take time out to watch Gary Yourofsky's inspirational speech on YouTube[16] or Earthlings.[17] I don't mean to just watch the first three minutes, walk away in disgust, and try to put it behind us, but to actually watch it from beginning to end and ponder deeply over the role all of us have unwittingly played through sponsoring such action via the goods we have bought as unwitting consumers. Of course, the ultimate vision is that there should be no such thing as a slaughterhouse, or an Earthlings video, but until that day comes, it would bode us well to witness the reality and to reflect deeply upon it.

Tolstoy Story

I am reminded of a little story I once read somewhere about Tolstoy, the Russian author of War and Peace, a renowned vegetarian of his era.

Apparently, his aunt once complained consistently throughout a meal he had prepared for her, due to the blatant absence of any meat to sink her teeth into.

Tolstoy, a man of many thoughts, pondered over this dilemma, and on a subsequent occasion of cooking for her, she was pleasantly surprised to be informed that she would be served chicken.

His aunt, no doubt gleefully anticipating her meal, arrived at the table only to find a live chicken tied to the leg of her chair, and a butcher's knife waiting on the table. Tolstoy answered her fluster by stating, "All you have to do is slaughter the animal, and we will be glad to prepare it for you."

The story goes that, being also vegetarian by nature though clearly ignorant of the fact, the aunt was unable to end the chicken's life, and from that point on, ceased eating meat.[18]

Thought Experiment

Test yourself. Picture this. You are hungry, and before you is a live chicken and a very sharp knife. Are you capable/willing to take the life of the chicken yourself? If yes, could you truly do so without riddling your conscience with guilt or, later or immediate, feelings of sadness, regret, or remorse?

I'm not too naïve to realize that there are some people out there who would stand up and answer yes, that they could do so, even gladly, but what I adamantly refuse to believe is that they would be being totally honest with themselves about not feeling the slightest sentiment of regret.

Avoid muddying the issue, and don't think of it as a do-or-die situation! Imagine you have a fridge full of vegan food.

Flesh Eating Is a Cowardly Practice

In my opinion, flesh eating is generally a cowardly practice. I say this because, like Tolstoy's aunt, few of us are genuinely prepared to personally confront the stone-cold bloodiness of murder.

We get others to do the work that we ourselves are unable to do, because in our hearts we all know how ignoble an act it is to deprive another of its life. Even the slaughterhouse workers are not ignorant of the ignobility of murder. Witnessing endless carnage day after mind-numbing day is utterly devastating to one's mental well-being, and frequently results in post-traumatic stress disorder. Such a "job" will inevitably drain one's physical, emotional, mental, and spiritual well-being. To be constantly surrounded by the animals' heart wrenching cries for help, as well as their blood, urine, and feces, must pay its toll. Desensitization is pretty inevitable for anyone who can take the job over a longer period, and purportedly also results in unprovoked episodes of rage and anger towards others.[19]

There is a reason we shield our children from the murderous facts behind their meals. We know they have yet to be corrupted and, should they be faced with the truth, will recognize the disturbing blood lust of life taking.[20] They have yet to be desensitized.

I recall a young girl I used to babysit back in the 1980s. She once witnessed her father remove a transparent plastic bag from the freezer. Its contents were a mangled portion of a lamb's carcass. After a while, as the bone and flesh began to thaw, blood settled at the bottom of the bag. Three-year-old Claire looked at it and said in all innocence, "Is that lamb's juice?" Her father looked at her and calmly answered, "No, my love; that's lamb's blood." Claire was silent for a brief moment, then with all conviction, replied, "I don't want to ever eat that again."

Of course, within a week her father had rationally and patiently convinced his daughter that it was perfectly normal and quite necessary to eat meat, and thus helped her on her path of increasing desensitization.

While our inner temperaments may be truly divine, our souls are constantly in a state of tumultuous unrest due to the atrocities our lifestyles help support and perpetuate.

Golden Buddha

I'm reminded of another apparently true story I heard once. It transpired some decades ago in Bangkok, Thailand.

The government had decided to build a major road through a certain area of the city. Unfortunately, it just so happened that a pre-eminent Buddhist monastery stood right in the path of the proposed highway. A relocation was thus in order.

Now it also happened that within the walls of this monastery, there stood a giant clay statue of a rounded Buddha that required relocation.

Initial attempts to lift the statue with a crane resulted in cracks appearing on the side of the Buddha. Not desiring to damage the statue further, the attempt was abandoned as they paused to reconsider their options. That night, it began to rain and, fearing the rain would widen the crack, one monk set about covering the statue. During his efforts to protect the effigy, he suddenly noticed something glittering from within the crack. Curiosity getting the better of him, he cautiously widened the crack to discover gold. Widening the crack yet further, he realized that the whole statue was solid gold with just a surface layer of clay.

Historians supposed that, at some point in time, monks had covered the statue in clay to hide it from invading foreign forces and that they themselves likely all fell victim to those forces, thus the secret of the Buddha was lost until that fateful day, centuries later, when the government decided their city needed a bypass.

This is how I picture us all to be. Born into the dementia of this asylum, the inner golden purity of our souls becomes gradually and increasingly more encrusted by negativity and dirt, until our shining true qualities are so suppressed and forgotten that we can no longer even recognize ourselves. If only a simple intercity bypass would suffice to make us realize the truth!

Hidden Industry

Shortly before starting on the writing of this book, while exercising my innate nomadic nature, I journeyed through Southeast Asia. There I witnessed skinned cats and dogs hanging from hooks, and stall after stall of dying and dead fish, eels, and chickens, their carcasses fly-infested and rotting in the tropical heat, and people carrying chickens tied together, hanging upside down by their feet, still alive, but only barely— sights that have long passed from being commonplace within the traditional western market.

It dawned on me just how much the animal exploitation industries in the west have become increasingly more concealed.

I believe this increasingly gradual concealment has been a natural consequence of the equally increasing rise in empathy and vegetarianism throughout the western world and has been, in essence, a necessity for the animal abusers, as they are fully aware that should the grizzly facts become open and publicly visible, evermore people would choose to not eat the flesh of other beings.

Thus, the more sensitive the masses become, the more the meat industry is forced to hide its bloody goriness, choosing instead tactics like cleverly presenting their wares on neutral, neatly packaged, sterile polystyrene trays, wrapped in clear plastic.

Even the words we use to describe the flesh we eat veils the frank bloody facts; instead of eating cows, we eat beef. Instead of eating calves, we eat veal; instead of eating pigs, we eat ham, pork, and bacon; instead of sheep, we eat mutton. Curiously enough, etymologically, many of these words come from the French words for the animals themselves: veau/calf/veal, boeuf/ox/bullock/beef, mouton/sheep/mutton, porc/pig/swine/pork.

Then there are those who, in futile attempts to appease vegetarian thoughts, console themselves with the notion that should an animal be murdered the right way, with the right ritual performed, and a prayer said afterwards, that the animal will actually benefit from being slaughtered as its soul will be given the opportunity to rise to planes hitherto unobtainable for it.

Beside the fact that I've yet to actually meet someone who, with all knowledge and forethought, genuinely consistently seeks out such slaughtered flesh, in my eyes this is naught more than another clever ploy to dampen the growing concern amidst the all too gullible masses.

There Is No Right Way to Murder Anyone

Look at it this way, if I invite you to live at our place, let you eat freely from our trees, clothe you, and give you comfort and companionship for several years, and then one day call you inside and shoot you, butcher you up, and eat your rump with scrambled eggs, then would it suddenly make everything hunky-dory if I went around saying, "It's okay you know; I fully respected the dude; I treated him well; he had a good time here. Shucks, I even listened to him while he talked of his philosophy on life and showed me his boring vacation photos, and don't forget he ate whatever fruit he wanted from my trees. I even blessed his body after slaughtering him and said a little prayer so his spirit could move on to higher plains! Rest assured, I killed him with love!"

Let's face the facts squarely and not try to hide behind some notion of ceremony or ritual; the reason I invited you was never out of respect for you—it was to fatten you up and shank you. When I saw you, it was not as a friend or an equal, or even as another being—it was as part of a devilishly greasy three-course meal, and when I made sure the wicked wolf didn't get you, it was only so I could instead.

One conclusion that can be drawn from traditional tribal ritual performed upon or after the killing of an animal is that those participating recognize that the act of taking another's life is not something to be taken lightly and that there is acknowledgement that something is not right with it.

No matter how you look at it though, murder is murder, and regardless of what prayers are said, murder is always committed without the consent of the victim. Hiding behind the unshakably, unknowable destiny of another's soul is a poor excuse at the best of times, and would likely not appease your own theoretical untimely assassination even if someone were to assure you that your soul would benefit from taking a shot to the head. Stop kidding yourself!

Slavery

Bear in mind the following:

A century and a half ago (not even), it was considered a reasonable practice by many, in the west and elsewhere, to forcibly capture people from certain regions of the globe and to thereafter freight them elsewhere, whereby they would slowly be beaten into accepting their fates as slaves.

People were, as the saying goes, treated like animals. But let's face it, not even animals deserve to be "treated like animals" in this fashion. Now 150 years further down the track, if we look back at those happenings at all, we probably do so with bafflement and grief that our ancestors could so easily have gone about their daily affairs, discussing the weather, while other people not too far down the road were being beaten and tortured into accepting lifetimes completely not of their own choosing. Meanwhile still, yet other fore-parents were actually trading in stocks of freighted people. Fine young specimens!

Who among us is able to put their hand up and say they honestly believe that what we were doing back then was right and just, and would be happy to rekindle those old habits? Sure, there may well be a few diehard triple K members still floating around out there who regret that the trade ever lost legal status,[21] but these shouldn't seriously be considered good role models for the masses. Besides, although mostly illegal, a multi-billion dollar slave industry is still very much in existence today. In fact, there are said to be more people enslaved now than ever before in history.[22]

The difference between then and now, however, is that the current slave trade is vastly invisible and mostly takes place without the consent of current global legal systems.

It is one of my profound beliefs that there will come a day, hopefully not too far off, when we will all look back at the relationship we currently have with animals, and be sick to our teeth about how we have used and treated them. Or, more appropriately, abused and mistreated them.

How we have deprived them of their freedoms, their natural environments, their families. How we have fed them grossly unnatural diets, and regularly restricted their movements to tiny booths, stalls, and cages, where they have been unable to even turn around.

All for nothing less than the sake of gluttony, hearing aids, and heart attacks, side dishes to our cheese and ham sandwiches (if that makes no sense to you now, as I'm aware I'm getting a little ahead of myself here, then hopefully it will become clearer as you read on through the pages of this book).

In Defense of the Fish

I have met many people who have taken first steps toward vegetarianism but continue the struggle to justify feasting on fish flesh.

There is no doubt in my mind that fish too are individuals, struggling to stay alive and unharmed, with as much aversion toward pain and captivity as any other animal species.

What a horror it must be to be caught up and cramped in nets, and forced to slowly suffocate in the air, or feel that yanking realization that the morsel of food we just ate has an embedded hook within it.

A Fishy Story

Imagine you are of a slightly different intelligence, walking along some street, happily anticipating lunch and just generally minding your own business, when suddenly you come across a slice of pizza, seemingly suspended in mid-air by some unseen force.

You sniff the pizza and your demented sense of smell tells you it smells good. You take a bite. Yeah. Good stuff! Soon you've devoured the whole slice and are wondering where the next one will appear.

But there is an annoying hair on your tongue. You try to pull at it, and suddenly there is a sharp pain from deep within your throat. This is shortly followed by an additional unexpected "tug," and the next thing you know, you cough up a mouthful of blood and your body is yanked irresistibly upward.

Meanwhile, up on the eighth floor of the block beside you, some geezer, cigarette drooping from his lips, baseball cap on backwards, leaning back on his stool, is reading his newspaper. He sees a movement on his line, folds up his newspaper, tucks it under the one leg of his stool, and starts reeling in.

Horrific what? I suppose you can be quite sure that the average cod with the average chips has not been caught in this fashion, but more likely has been trapped by a net and pulled out of the water with thousands or more of its brethren and left to steadily suffocate. Flapping around spasmodically, slowly, and painfully asphyxiating.

Either way you look at it, it's not too nice a scene. But let's face it; it's truthful. Those who call fishing a "sport," have to realize that there is nothing remotely "cricket" about it.

Yes, kid yourself no longer that you need to eat animals to live! This is an absurd notion. This is what is meant by the expression believe absurdities and you will commit atrocities. It does not have to be this way!

Carnage Statistics

Human lust for flesh-based foods causes the murder of literally tens of billions of animals annually. Roughly sixty billion land-based animals alone are killed yearly. Factor in also water-living creatures, and the figure soars to over a hundred billion! That's a shocking 100,000,000,000 animals whose lives are cut short purely for human gluttony.[23] The US alone is responsible for more than ten percent of that number, with twenty three million chickens and one hundred thousand cows (over four thousand every hour) killed daily![24] These figures are surely a conservative estimation, and by some reckoning, the massacre toll is said to be even higher. One source, for example, states that in 2011, including shellfish, sixty three billion animals lost their lives to feed Americans alone.[25] ADAPTT gives a very disturbing real-time counter of globally slain animals on their website.[26] The truly accurate kill statistics are beyond counting, but to be sure, the majority of victims suffer far more intensely than most homicide victims. These are murders you'll not find mentioned in the daily tabloids. It's time for us all to face these facts and acknowledge the unwitting role we've all played in shaping them.

Flesh Alternatives

If you are perchance currently a shod omnivore, and I've somehow managed to reach your better sensibilities, leaving you with desire to take this first step on the path to Eden, but are desperately craving steak like food, then be aware that the market is flooded with alternatives to ease one over to the vegetarian diet. Many are bean-based, and can be prepared and eaten in much the same way as animal-derived sinews and thews; you can even buy rashes of soy bacon and mince-meat-similar TVP (textured vegetable protein). Scientists (bless their precious cotton socks!) have now succeeded in fully synthesizing meat in a laboratory.[27] And the very latest news, hot off the presses, is that bioprinters may soon be able to print meat![28] There is NO valid reason to continue supporting the flesh trade.

Chapter 3
Veganism

Now, although becoming vegetarian is surely a good step toward bringing more peace to this planet, as an ethical dietary choice, it is somewhat lacking and rather hypocritical. Hence veganism steps in to iron out the flaws and crinkles of vegetarianism.

Defining Veganism

I've been wanting to write this for some time now. I want to make it very clear from the onset that veganism, despite frequent contrary claims, is not and never has been a diet.

So, you know how some people say that they are vegan for health reasons? Or vegan for environmental reasons? Or even vegan for social reasons?

Well, technically speaking, they're all wrong!

They are wrong, that is, unless they have redefined the term "veganism" from what was originally intended. Which of course, is what has colloquially happened to many words that have been misused and abused, as is the plight of the daily spoken English language. So, on second thought, perhaps that makes them right in a sense.[29] But not if one takes the literal meaning of the word as it was first coined back in the 1940s.

I believe the original definition, as per The Vegan Society goes like this: "Veganism is a way of living which seeks to exclude, as far as is possible and practicable, all forms of exploitation of, and cruelty to, animals for food, clothing or any other purpose."[30]

If you went to the source of that quote, you might have noticed that I deliberately left out the part about how, in dietary terms, veganism is generally further defined as the practice of dispensing with all animal produce, including meat, fish, poultry, eggs, milks, honey, and their derivatives, but only because the real golden rule is to leave animals be!

I feel this is worth emphasizing because too many people focus on veganism as a diet, failing to recognize that the only real guidelines it demands of one's meals are that no animal should be deliberately and willfully harmed in its production. I think we can all agree, for example, that road kill may be completely and utterly void of any exploitation of or cruelty toward the victim animal. Thus, personally, I would find it difficult to see any real moral issue preventing the carcass from being used in any way, including as a potential recipe ingredient.

I make this intentionally provocative statement not by any means as a proposition, or recommendation that we begin scavenging for our food on the sides of roads; I merely wish to make the point that there would clearly be nothing inherently contradictory to the uncomplicated veganism definition above: a way of living which seeks to exclude, as far as is possible and practicable, all forms of exploitation of, and cruelty to, animals for food, clothing or any other purpose.

Personally, with the exception of one rather eccentric Paleo acquaintance I don't think I know of anybody who would seriously entertain the idea of road kill casserole; certainly I've never met a vegan that would! I would imagine that most would find the same kind of repulsiveness to the proposition as they would when faced with the proposition of eating their own naturally dead grandmother, so please don't try to take me too literally here folks. I'm merely trying to get you to think outside the box; in my mind, scenarios such as these further underline the fact that we are clearly not an omnivorous species; every real omnivore out there would have no trouble in recognizing a free meal!

Veganism is a lifestyle choice that attempts to minimize undue suffering toward animals; why would anyone consciously choose to harm an animal when there are other things available to eat, drink, and wear?

Real Men Don't Eat Quiche

When you think about it, vegetarianism, or veg"eat-dairy"anism as one friend once so wittily put it, is a very human diet, at least as it is put into practice by the majority of adherents out there. (Note: that was human diet, not to be confused with humane.)

What I mean by this is that no other animal species consciously chooses to not eat the flesh of another while still imprisoning winged beings to systematically steal their eggs to fry, ending their lives prematurely after egg-yield begins to diminish, and does likewise with large mammals, raising them until child-bearing age, systematically raping them and forcefully separating mother and child in order to carton off the milk that is intended for the offspring, and send it to the waiting dairy-crazed milky masses. White with two sugars?

And further, a human will happily go out and buy the flayed skin of a slaughtered innocent, skillfully fashioned into footwear, trousers, skirts, belts, jackets, gloves, and handbags.

Again, I'd like to clearly underline that the true essence of veganism is not about diet, and thus offers no clear guidelines as to how the diet of a follower should or shouldn't be.

There is no underlying concern if the diet is raw or cooked, with condiments or plain, mixed or mono, toxic or palatable, rotten, ripe or unripe, deep fried, roasted, boiled, sautéed, smoked, alcohol fortified, preserved, MSG flavor-enhanced, colored, fermented, pasteurized, curried, sugared, spiced, or just plain natural—provided of course, no animal was exploited during its production!

My point is that veganism has little outlying regard or concern for the health of those who practice it. Don't get me wrong, I am not saying that vegans aren't concerned for their own health; of course they are, who really isn't?

Who doesn't want to live a long, prosperous and contented, healthy life?

All I am saying is that those claiming to be vegan for health reasons do so not for concerns stemming from the philosophy of veganism or any of its original guidelines, but from their own will to live happily and healthily.

Actually, the English language is really missing a word to describe someone who purely eats a diet free of animal products. The French for example, in addition to the word véganisme, have the word végetalisme, which describes the philosophy of just eating solely from the vegetable kingdom. So they have végétarisme, végetalisme, and véganisme. Similarly, the Spaniards do the same, with the words vegetarianismo, vegetalismo and veganismo. I guess the English equivalent of vegetalismo would have to be dietary veganism.

So, put simply, veganism is an attempt to avoid being the cause of unnecessary animal suffering.

Dairy Cows

· I've met so many vegetarians; some are extremely sensitive individuals who knew instinctively from very young ages that eating the flesh from someone else's body just wasn't cricket.

Surprisingly though, many of them are very reluctant to hear the truth about dairy. In their hearts, they must see that something is wrong with it, but their stomachs really don't want to listen. They conveniently prefer to remain totally ignorant of the facts. Cheese, yogurt, and eggs are, in many ways, just as addictive and harmful as (or perhaps even more so) than flesh products.

Actually, it is my firm belief that, in reality, production of these items may be causing more cruelty and suffering to our fellow brethren than the trade of their flesh does. It may, for example, be argued that flesh can conceivably be taken from bodies whose original owners have lived freely in their natural habitats until one fateful day when a bullet retires them early.

On the other hand, the same can never be said of milk products. Please don't be misled to believe otherwise. Animals reared for the purpose of milk production are only permitted to live while they are able to produce large enough quantities of milk to satisfy the financial desires of the producers. Naturally this is only until the biological end of the animals' peak offspring rearing years. For cows this means having forced pregnancies from as early an age as possible, each year, for roughly five years, at which point they are deemed no longer ideal and thus are slaughtered.

No other animal species takes milk naturally after it has left the stage of dependency on its mother.

Yet, somehow, the mass marketing, advertising industry has completely hoodwinked the majority of us into believing we need milk all our lives and that should we miss out, we will undoubtedly suffer from the lack of it.

Still Breast Feeding at Ten Years Old

Hmmpf!

Examine your feelings when you hear about the hypothetical woman still breastfeeding her own child at ten years old. Now go and open your fridge! Are you telling vegans to grow up and get a life? Open your fridge! Grow up and wean yourself!

The molecular constituency of cow's milk is so radically different from that of a human mother's milk, that calves quadruple their weight within the first months of their lives. Ponder over the origins of childhood obesity!

Cows are pumped full of hormones and antibiotics in order to increase milk yield and stop infections, mastitis, etc. Many of these chemicals are transferred via the host into the milk. Milk after weaning[31] is highly unnatural for us and any other animal species, milk from another species even more so, and pasteurized hormone-saturated milk from another species even further so.

The average modern dairy cow regularly produces between thirty-five to fifty liters a day.[32] Recently, one cow has been reported to be producing a staggering one hundred liters of milk a day.[33] Imagine the cow's udders dragging on the dirty floor, and, to make matters worse, her back would surely cripple.

Chickens

Natural chickens running wild in their native non-interfered settings have been observed to produce an average of six to twelve eggs every few months, all of which, generally speaking, hatch into fluffy young chicks.

Chickens on farms, however, have been known to lay in excess of 300 eggs a year, a feat that must have a toll on the mother's poor enslaved body!

A single battery farm can contain hundreds of thousands of battery hens, stacked in cages five rows high in enclosed sheds. These hens suffer intensely and continuously throughout their short, miserable lives. A battery hen may spend her entire life lying cramped inside a shared cage, each containing three to twenty hens. Each hen's floor area is equivalent to little more than one A4 sheet of paper.[34] With her wings outstretched, a hen is twice the size of a typical battery cage.[35]

Knowing these facts, some people might choose to eat free-range or organic chickens and eggs, believing that they had a fair deal. Sadly, they are misinformed. To increase profits, many free-range (and organic!) farms cram literally thousands of their chickens on mud-filled lots or together in sheds with a simple access door to the outside that only the closest chickens get to use.[36] These animals often suffer through the same mutilations (debeaking), without painkillers, that occur on factory farms. Actually by some accounts, due to the extreme overcrowding and filthy conditions, organically raised chickens sometimes suffer from higher mortality rates than their drugged battery farmed equivalents and, due to the lack of antibiotics, can end up with even more parasites than are found in factory chickens.

Each one of those chickens has its own thoughts, its own unique personality and individuality, experiencing little more than endless discomfort and merciless unremitting agony.

Leather and Cheese

The hide of cattle can sometimes make up ten percent of the animal's gross financial value. Kid yourself no longer that leather is just a side product of the industry.[37] It is as much part and parcel of the animal's ultimate suffering, as the trade in its flesh and the grinding of its bones for gelatin is.

Due to the rennet ingredient, taken from the stomach of the slaughtered animal (often from the slaughtered young), the vast majority of cheeses should not even be considered vegetarian.

Wool and Silk

I think it's pretty obvious how fur farms work,[38] so I see no reason to write more on that. But one animal covering, which is often little understood, is wool.[39] Many may erroneously believe that it's removed without harm to the sheep, but the wool industry is yet another example of animal exploitation. The sheep (and in some cases goats and rabbits[40]) are bred and valued for one main purpose—their wool. When the wool diminishes, they too are sent on a one-way trip to a slaughterhouse. To my knowledge, shearers are never paid by the hour, but, rather, by the number of sheep they shear, thus the incentive is to get through each sheep as quickly as possible, which results in very rough handling. There are more issues to be aware of too.[41]

Silk worms are bred and killed by the hundreds of millions yearly. Thrown into boiling water, roughly fifteen are needed to produce just one gram of silk![42]

The Best Kept Cows

To make clearer the plight of even the best kept cows, I'd like to tell you the following short story. Although told from the viewpoint of a cow (I might add one of the luckier ones who avoided the factory farm), it is actually a very true story, based on actual occurrences. The one man in the story referred to as "the Gentle-one" is completely non-fictional and it is through his recounting of events that I penned this story.

My Life as a Cow

Spring had once more descended upon us, and for the second time during my brief existence, I found myself heavy with child.

Birds chorused their spring melodies and the wind bore its sweet, fresh fragrance. The only evidence remaining of the cold winter were the odd patches of snow framed by silver linings of melting ice at which the sun slowly etched.

We had seen none of it though. Winter has been a season we were never given the chance to know, imprisoned indoors and confined to small booths. But with the commencement of spring, we were once more ushered outside to the waiting grass, sun, and above all, renewed hope.

Within me, I felt my child's restricted movements, its ever-increasing impatience, gently but forcefully informing me that birth was imminent. With encouragement, I sent out feelings of love—to wait, to be patient. Not yet, the time was not quite right.

My thoughts, without words but thoughts nonetheless, retreated to the events of the previous year. Spring had barely begun, and after just a very fleeting moment of contact, my first-born had been forcibly separated from me. Who knows what fate awaited her. I mourned the loss for many weeks, searching for her, calling desperately. But to no avail; never did I see her again.

But they were happenings of the year past, and although I still remembered (indeed, how could I possibly have forgotten?), time had slowly healed over the wounds of this traumatic loss. Make no doubts, this is a hard and cruel life I have led. Only the simple pleasures have kept me sane throughout it all. I did not suffer alone. My sisters offered warmth and gentleness, though their circumstances were clearly no better than my own. They too have suffered the loss of their children at the merciless hands of humans. But I would not give up hope. I could not.

During the autumn, before the air turned cold and we were moved permanently indoors, the new one arrived. The Gentle-one. Unlike the Rough-handed-one, his voice did not bellow, he showed signs of kindness, and contained reassuring words of comfort. He would visit us daily, offering soft words and firm neck strokes. He attached the plastic tubes to the teats of my udder with more care and consideration. Though truthfully the milking was always more of a frustrating, cold ordeal—never a pleasing experience—and I was always glad when this was over. Please don't underestimate my intelligence; I know what I have missed and for whom my milk was truly intended.

Yes, I held hope that the Gentle-one could help. I believed he would understand, and as the time drew close, I silently left my sisters and headed toward the stream where I could see him working in a neighboring field.

"Mmoo."

"Oh, it's you, Emily. You gave me a fright. What are you doing down here? You know you should be in the other field with the rest of the herd. Come along now; get back with you."

Of course, his words meant little to me, but his intention was clear; he wanted me to return. He pushed my head gently backwards and slapped my side firmly, encouraging me to leave. However, his voice seemed neither harsh, nor reprimanding, so I stood my ground firmly, aware that the birth was only minutes away.

"Come along now, Ems; don't be a silly girl. I don't have all day; there's work to be done here, can't you see? Best you be getting back now."

There were more friendly words and gentle attempts at persuasion, but he didn't seem to mind my presence too much, so I stood firm and waited.

"Okay, stay there if you wish, but don't come too close, or you'll be liable to hurt yourself."

He turned once more to his work and whistled contentedly and, I believed, soon forgot about my presence. After a while, I felt movement within me. Very quickly and smoothly, my son entered the world. Bloody and small and helpless. I licked his face, clearing a passage for air to enter his nostrils and mouth, and cleaned around his eyes and ears. My son, no one could deny the love bond between us. I smelt him, cleaned him, loved him, and as our bodies moved closer, he found my udder and took his first taste of its rich milk. This is what I had been missing—this gentle creature's tongue at my udder. I let out a "Mooo" of ecstasy.

"Yes, hallo Emily. Well, what have we here? A mighty fine looking young beef calf. Well done; good work Emily. He'll fetch a good price. I suppose we'd best be getting you both back."

He moved over to my son and, hoisting him over his shoulders, moved off in the direction of the rest of the herd. Naturally my son was alarmed, but I mooed reassuring moos and licked him comfortingly on his neck as I followed closely beside them both. I felt quite sure that things would be different this time around.

He put my son down and left us both standing with my sisters. Three full days my second child and I spent together. On the fourth day, two men entered the field at dawn and separated us. That afternoon, I found myself once more in a booth in the milking parlor, with tubes attached to my udder.

Severely stressed and frustrated, although the Gentle-one's voice was just as firm and soft as always, there was no more reassurance, and waves of confusion swept over me. I called out and cried for several days, often long into the night, but my cries were in vain and eventually I gave up hope of ever seeing him again. Routine life with my sisters continued and we were all able to comfort each other to an extent, as we had all suffered the same loss.

Summer arrived, and once more I found myself fastened to the rape rack. A bull, of jersey stock, brown, white, and oversized, mounted me, heavily, clumsily planting his seed within me. Though I hasten to add, I never felt, or feel, any resentment toward the bull. How could I? His lot was surely similar to that of my sisters and mine. All I felt was frustrated pity.

The daily milking continued, and the touch of both humans brought forth feelings of anger and frustrated loss.

Eventually, the seasons changed once more, the winds came carrying cold air, and someone saw fit to move us indoors again, keeping us locked in our small booths. Unless you have lived in one yourself for weeks/months on end, you can never understand what an experience this is. The days blend together and only the strange lighting and regular milking hours offer hints of the movement of the sun and sky outside. Those days indoors were truly the most enduring and perplexing.

For the third time in my short existence, I felt the growth of life slowly taking form within me. This only succeeded in increasing my stress. I had to be outside when it was born, to hide my child somewhere. Yes, even back then I had the inklings of a plan forming within my mind. Naturally there was no desire to have my child stolen from me for a third time.

Spring arrived and we were lead outside. There was no sign of the winter having visited. No telltale patches of snow. Perhaps the winter had been mild, or perhaps our stay indoors had been longer.

We kicked our legs and breathed the clean air. It was so good to be outside once more. Though, as we were soon to discover, our roaming freedom had once again been restricted. An electric fence had been raised, cutting off the field to the west with the stream.

Several weeks later, the hour of birth drew nearer—maybe tomorrow, I thought, as I sat down to sleep in the field. It was good to sleep outside, to hear the wind and the insects and the gentle breathing of my sisters. Not now, little one. Heavy with anticipation, I sent out feelings of love and encouragement to the child within me.

The following day, feeling that the time was close and seeing a chance, I slipped away from the rest of the herd, heading for the woods. After searching around for a while, I eventually found a suitable spot. No one had seen me. I gave birth to a boy, beautiful and strong, but small and fearful. I licked his face clean and offered him my udder, but feared to stay longer lest someone notice my absence. Alone, I returned to the field.

Later that day, the two humans came visiting.

"Looks like Em's dropped her calf, Frank."

"Yep, sure does; can't see it anywhere around though. No doubt she's hidden it somewhere. They do that on occasion. Crafty little buggers. Keep a close eye on her. No doubt she'll return to it later on."

"Sure will. Where'd you hide it eh, Em? In the woods maybe? Huh? I'll go and see to the chickens, then get back to work on the tractor."

"Okay, I'll be loading up the pigs ready for the sales this afternoon. See you later on."

I didn't know if they suspected anything or not. Perhaps they weren't even aware that I had been expecting. How was I to know how much they knew? I chewed grass with my sisters and tried to be calm.

Nighttime and darkness descended upon us, and only then did I dare leave the herd and venture towards my newborn son. I found him where I'd left him, sitting there, calling out to me. I nuzzled my head reassuringly against his neck. He drank thirstily from my udder, no doubt frightened and confused as to where I had been and why I had abandoned him. I licked him and moved my body closer to offer my warmth. But then, something wasn't right, there was movement behind me. A sister maybe? I turned around, and by the light of the big round moon, saw two figures approaching through the trees.

"Here they are, Pete! Knew they'd be here some place. Not the first time this has happened. Little bastard bringing us out in the middle of the night. You grab the calf and I'll get Emily back to the field."

"Mooooo."

"Okay. Looks like we got ourselves another future few hundred kilos of beefsteak. Will fetch a good price at the market, this one. Come along now, little one. We won't hurt you."

"Mooo."

What was happening? The Gentle-one had lifted my child over his shoulders and was carrying him away from me. I tried to follow, but the Rough-handed-one barred my way and urged me roughly back in the direction of the field. My son cried out for me and, in panic and confusion, I leapt forward and around, away from the Rough-handed-one, scraping myself clumsily against branches. Eventually I worked myself free of the woods, but by then they were already halfway across the field. My son, frightened, called out to me. Run as I might, I was unable to reach them before the gate closed, barring my way.

I watched the Gentle-one enter the buildings with my son. I cried, I mooed; oh believe me, these were traumatic experiences. Have you ever had your child removed from you and were not able to do a thing about it? My experience was nonetheless painful without the words to describe it.

Weeks passed, and I neither saw nor heard my son again. What kind of life is this I have lead? Why? What have I done? In my own way, I have asked myself these questions, but can find no answers.

Four children I have had in this short lifetime. Each one of them has been stolen from me. The regular irritation of plastic tubes against my udder has resulted in sickness and infection on many occasions. My teats are constantly sore, and inside I have stomach pains. I feel old although I have only experienced six summers. My milk has run dry and now, after having been driven for several hours in a hot and badly ventilated vehicle together with many of my sisters, I stand inside a big white building that reeks of anxiety and death. One of my sisters fainted during the journey, and I watched with horror as a human beat her to regain consciousness and forced her to stand up and move. I have also been beaten with a metal rod, and have a damaged hind leg. I am barely able to stand erect. I am afraid and I know what is happening here—my intelligence is enough to realize. I can see the stains on the floor and on the aprons of the humans. I can hear distant sisters calling out in such frighteningly horrible manners in neighboring rooms. I have never heard such intense, distressed cries. I stand stupefied and await my fate in dread.[43]

---

*It should be noted that although this story may be touching for many, the fate of most dairy cattle is in reality far worse. Most of them are completely unaware of what really goes on and how unbearable the life of a dairy cow truly is. At roughly two years of age, young female cows are forcibly impregnated. Within twenty-four hours of giving birth, their calves are removed and usually slaughtered within the first five days. The mother is left grieving for her lost calf, whilst her milk, which nature intended for her baby, is pumped, pasteurized and sent to supermarket shelves.

Roughly seven weeks later, she will once more be forced pregnant, and this cycle will continue until she is no longer able to produce profitable milk quantities. At this point she will herself be murdered, butchered, cooked, and gluttonously devoured.

In Defense of the Bees

Honey is one particular issue that some vegans have disagreements on. The overly sweet nectar syrup has captured the taste buds of many, blunting their empathy for the plight of bees.

As with pretty much every other ethical issue though, it is normal that humankind attempts to justify its behavior, and it is often reasoned that our relationship with bees is more of a symbiotic one. We create for them the comfort and security of their hives, and in exchange, the honey is extracted. Anyways, the bees will always produce more honey than they themselves actually require.

Without bees, this planet would likely be in dire straits. They work hard for their honey and deserve every drop of it. The reality is, though, that their hives are systematically, seasonally ransacked, and often with buckets of cheap sugar water left by the side of the hive as a poor substitute.[44] To do this, the bees are often smoked out to make them docile, before their homes are invaded and their honey is removed.

Imagine that you have been working for months, harvesting your own food, and then someone comes along and steals it from you. You'd get angry, right? You may accuse me of anthropomorphizing again, but this is precisely what every other animal will do if their home and livelihood is under threat. They will attempt to defend their home and their well-earned food. Any attempt at defense with humans as the aggressor, rarely ends in success. Watch those bees and see if you can find a better emotion than anger to define their moods.

Of course, there is no denying that their homes have indeed been crafted specially for them, but everything comes with a price and, in this case, the cost is far more than just the stolen honey. The bees are, after all, unaware of having signed any contract!

The bee populations around the world are currently dying in unprecedented proportions. It is generally suspected that this has to do with the ever-increasing use of pesticides (such as the Monsanto flag-ship product, Roundup) and I have no doubt that is partially to blame, but the long-term consequences of living in their artificially constructed homes and having their natural food source systematically pillaged from them, often in exchange for something nutritionally inferior, is also playing a major role. It's making the species progressively weaker, more intolerant of fluctuations in temperature, and more prone and susceptible to disease of many kinds. Back when they were more of an independent insect species, they would build their hives in the hollows of trees. Sure, it involved more work for them, but it was the work that nature intended for them and, as such, could only be of benefit to them. Such natural hives would have experienced more ranges in temperature, rendering the bee hardier than today's council-flatted equivalent.

Being an observant gardener, I too have noticed that there are far fewer bees than there ever used to be, and it is certainly disconcerting to imagine what may happen should the species dwindle yet further. Their role in nature is so crucial to the life of plants that without them the world would be at a loss. Their role needs to be acknowledged and respected, and, ethically, their honey should only really be considered as food should the bees themselves abandon their homes in search of further life adventures.

Leave Us Alone!

Shod omnivores and fundamentalist veggie-phobes who feel threatened by the vegan movement like to argue for their own rights to eat whatever they choose and to do so without finger pointing or interference. If it were all purely a question of food, they would have every right to protest at any finger pointing, however, they are effectively arguing for their right to deny the rights of others. Nobody is physically preventing them from eating the cow, but through their rapacious desires and consumer demands, they are fully denying livestock any rights, causing them untold harm and grievance and at the end of their restricted short tortured lives, cold blooded murder. And we should stop voicing our disapproval?

Off-the-shelf phrases like "to each their own" prove just how little thought people give to their arguments. The systematic torturing, slaughtering, and butchering of animals should not be compared with one's choices in fashion, film, music, or religion. The truth of the matter is that your freedom of choice to eat meat and other animal products is basically about paying someone else to commit carnage.

Protesters will likely argue that animal killing goes on all around us and that virtually every species participates. Yet, when it comes to human rights, they themselves will likely agree that murder, child molestation, rape, slavery, unwilling prostitution, and more should be policed and prosecuted. Despite this, there are likely examples of the above within the behavioral realms of other animal species. Although I agree that we can learn from watching wild animal interactions, I think it rather narrow-minded to blatantly assume that their lack of moral conduct justifies our own failing in that regard.

I'd like to point out that marital rape[45] was, and in many countries still is, considered "legal." In fact, it is only within the last fifty years that some countries have begun to criminalize such cases—England not until as late as the 1990s! Many meat eaters would likely object if a rapist were to defend their behavior as being purely a question of "personal choice" and that others should not impose their moral values on them. To refuse to debate an ethical issue on the grounds that your actions are no one else's business is to fundamentally misunderstand the nature of morality: "These self righteous anti-rapists are so full of themselves. Interfering shazbats should mind their own darned business!" Thus, another demonstration of a popular tactic employed by meat eaters in defense of their supposed freedom of choice: "curses" and "insults." It's always proof that you are winning an argument when you start to get sworn at!

Clearly what is currently legal or illegal is not always an indication of whether or not something is moral or immoral. So just a few years ago, when marital rape was still considered legally acceptable behavior, those who could see it as unethical were protesting with good cause. Fast-forward to today and a vegan, quite similarly, will voice his or her indignation about the unnecessary exploitation and slaughter of sentient beings, even though that too is most clearly fully within legal boundaries!

Everyone is entitled to their own personal tastes, free from interference, but—and this shouldn't need saying—justice is not a lifestyle choice.

It is also not a question of anyone being "better" than anyone else, another ill-conceived view that I've heard expressed on more than one occasion. The only person we should be trying to be better than is ourselves, yesterday.

Sense of Humor

I have read and heard, more than once, shod omnivores accusing vegans of having no sense of humor. This is a rather silly accusation based mostly at our unwillingness to laugh, and find funny, sick jokes concerning the suffering of others.

If a serial killer were to joke about his bloodily executed victims, would you seriously find yourself laughing whole-heartedly with him? Those poor jokes about the juiciness of a good steak need to be fully grokked. No, I'm not accusing shod omnivores of being serial killers, but I am accusing them of aiding and abetting serial killers in the continuation of their serial killing. Once you've fully grasped the facts, it will be clear just why talking of the deliciousness of bacon is not the least bit comical. Go back further, and that bacon was the flesh on the stomach and back of a living, breathing, sentient, individual being, who, born into misery, likely never knew anything but suffering and torment throughout its entire lifetime, and that life was macabrely cut short for no other reason than to satisfy the gluttony of shod omnivores.

Now laugh!

Father Patrick awoke one morning to discover a dead donkey on the front lawn of his house. After examining it and confirming its state, he rang the village police. Father Patrick awoke one morning to discover a dead donkey on the front lawn of his house. After examining it and confirming its state, he rang the village police.

Lisa: "I'm going to become a vegetarian."

Homer: "Does that mean you're not going to eat any pork?"

Lisa: "Yes."

Homer: "Bacon?"

Lisa: "Yes, Dad."

Homer: "Ham?"

Lisa: "Dad, all those meats come from the same animal."

Homer: "Right, Lisa. Some wonderful, magical animal!"

Yes, yes. I laughed at *The Simpsons*[46] joke too, but only because it was a shod omnivore joke, not an animal one! It seems many are content with Homersapienism as a valid next evolutionary choice. Okay, enough with the chiropractic mirth; sidesplitting, rib-tickling, ROFLing, and thigh-slapping aside, tough issues are raised here, meriting deep reflection and sober consideration.

Dairy Alternatives

So, want to become vegan but can't live without milk? Despair not— supermarkets are full of dairy alternatives like soy and rice milks, soy cheeses, even egg alternatives. Back when I first turned vegan in the 1980s, my wonderful sister and I discovered that flax seeds, boiled mildly in water, form into a gooey gluey, egg-like substance, and with further experimentation, we found it was a perfect substitute for a binder in cake recipes. Of course, I'm not saying any of these foods are ideal (especially not if GMO-based!), but at least you'll be able to look a cow and chicken in their eyes and have a clearer conscience. There is NO reason to continue purchasing animal products.

Chapter 4
Raw Veganism

Go back quarter of a century or less, and the word "vegan" was still virtually unheard of. Finding another vegan in your hometown was far from an easy task, and many of us felt quite isolated. In the 1980s, I had this small backpack with the slogan, "veganism is the way forward," written in bold white letters on it. You'd be surprised how many times people would stop and ask me what it meant. Nowadays, things have changed considerably. Nearly everyone in the western world has heard the term, and most people even know, or know of, a vegan or two. The movement has grown substantially, especially since the dawn of the Internet, and hooking up with fellow vegans is generally no longer a difficult task.

Over the past decade or so, the idea of not cooking one's food has also been on the increase, both within the vegan movement and within the health conscious omnivore populace. For many, the choice to do so is solely one of health and fitness; once one delves deeper into the idea, the health benefits of not cooking become pretty obvious. However, reasons to eat raw go far beyond matters of physical well-being. I intend to write more on this in following chapters.

It's true that the cooking process renders many foods edible that otherwise would be rejected raw. Many see this as a blessing, and indeed there may well have been past times of hardship when pockets of humans would not have survived without these foods. However, those times are over, and it's time we move on from such sustenance and begin to understand the full extent to which our diet shapes the world, and the ethical role it must inevitably play in preparing for a positive more love-filled future.

The ultimate goal of ethics is to refrain from all wrongdoing and harm. This is something all true vegans are aware of. It does not mean one cannot or should not break the law, especially not if the law is clearly unjust, but it does mean that we should try our utmost to not cause suffering. With this in mind, it should be clear that we have a duty to also educate ourselves and refrain from practices that harm our own physiological well-being. We must treat ourselves, our own bodies, as ethically and best we are able. It is our duty to guide and love ourselves, to understand that our bodies are not ours to abuse, but to maintain, nurture and cherish, and use as examples to sway and influence others positively.

For many, the steps from omnivore to vegetarian, and vegetarian to a vegan diet, are taken for precisely these reasons. For one's budding desire to improve one's health and consequently life, happiness, and longevity.

One's health is clearly a valid reason for making dietary changes, but unless we can see beyond personal well-being, the likelihood of failure is high. Purely egotistical reasons for change are sadly rare enough to ensure consistency, and often result in unnecessary and harmful compromises—eating a fish, for example, because one suddenly feels like it, or sharing the carcass of a turkey because it's Thanksgiving and one doesn't wish to upset family. There is no reason for anyone to be upset at another's statement of empathy! Lifelong commitment to dietary change and personal improvement are rarely successful without first broadening one's powers of empathy.

If you do the right thing by all parties concerned (i.e., the fish and the turkeys too), you will also be doing the right thing by yourself! If you wonder about doing the right thing by your host, realize that there is nothing ethically wrong with refusing to eat a meal. If someone gets irrationally upset with this choice, that is purely their issue. Don't let it become yours! If you are consistent with your stance, others will soon accept the decision you have made, even if they don't agree with it.

As mentioned, I intend to get back to ethical issues beyond the duty to one's corporal form, but first I need to explain why and how eating raw is physiologically the right choice.

Why Raw?

I'll begin by asking a simple question: what's the furthest you've ever ventured away from human civilization (or "sillyvization," as I prefer calling it, if you'll pardon the Bugs-bunnyism)?

I ask this because if you've ever had the opportunity to observe any wild animal, regardless of species, in its natural habitat, you will surely know that all such animals eat their food in its raw, natural state. Human beings and domesticated and feral animals living in close proximity of humans, such as urban foxes, are the only exceptions to this rule.

We, on the other hand, excel as omnivores, eating virtually anything from someone else's fried kidneys, to birds' nests, to piss-drenched eggs,[47] and chocolate-covered grasshoppers.[48] With human ingenuity, there's really little that cannot be rendered palatable.[49]

I expressed earlier that we should steer clear of looking up to animals as role models, and generally that's the stance I prefer to take, but in this particular case, I take exception. I really think that human ingenuity has gotten the better of us here. Instead of eating food as nature offers it, we harbor this perverse idea that food should be burnt before consumption, with some people going through life never even tasting anything the way nature intended. Not only has such habit become so ingrained, but we also go about telling everyone how healthy it is too, believing erroneously that we are doing ourselves an injustice if we don't get at least one cooked meal a day!

At this stage, some readers might be wondering where I am headed with this, and what logic there is behind me stating that such thoughts can be anything resembling erroneous. Bear with me a while, as I intend to get there.

First though, I'd like to point out what I believe to be obvious. Humans are probably the sickest, weakest, most neurotic species in the globe, not even inbred pit bull terriers coming a close second. We have more flavors of disease than one can possibly shake a stick at. Our bodies morph into deformity more and more each day, making us gradually lose our natural born beauty, such that over time we become unrecognizable from how we should look. This may not be clear to those of us who have never given this any thought, because this deformity I talk of is considered relatively "normal" by most, a natural consequence of life and ageing. This is so far from the truth.

In 2010, chronic diseases were the leading causes of death in Australia, with the most common causes including cardiovascular disease, cancer, chronic lower respiratory diseases, and diabetes.[50] More than anything else, it is my belief that such ailments are predominantly caused by diet and subjecting one's food to chemical mutating, life shattering heat prior to consumption. I will now attempt to make it easier to digest, in case none of this makes any sense to you.

Organic Food

How many of you go out of your way (or at least feel you perhaps should) to obtain organic fruits and vegetables?

It's bizarre, perhaps, if one should then return home with them and render them consequently inorganic.

Think about it. Organic also means pertaining to life, thus, removing the life force from within a food, as is the result of raising its internal and external temperatures above a certain degree for an extended period of time will consequently render any food, regardless of how it was grown, inorganic. Hmm, defeating the object somewhat maybe?

The Kitchen as a Chemical Laboratory

Whether we are aware of it or not, the conventional kitchen can be readily compared to a chemical laboratory. Applying heat to a substance is one of the primary methods used in altering its chemical makeup. The chemical structure of all substances changes radically under application of intense heat, a process that inevitably results in destroying all living cells within any food subjected to it for long enough. You don't have to stick your hand in a flame to understand what this means! It will no longer be the natural, energy-pulsating, nourishing, life-giving substance it once was.

I have no doubt that it is far healthier to eat conventionally-grown, chemically-sprayed foods, washed and in their raw natural uncooked states, than it is to eat spray-free, organic, or biodynamically grown foods cooked.

Picture two carrots, one of them organically grown and cooked, and the other a conventionally grown and thus chemically sprayed one, still raw. What do you think is the difference? Stick them both in the ground and observe. The cooked one will quickly rot, putrefy and grow sludge-like mold, whereas there is more than a reasonably good chance that the raw one, despite the chemicals, will live on, sprouting itself a new green leafy carrot top. Similarly, eat the raw one and it will transfer its life-giving force to the body; eat the cooked one and it will partially end up as internal gunk. The physiological effects of the chemicals created through the cooking process are far more proliferous, thus more harmful, than the conventional ones used while growing the food.

I'm not trying to convince anyone that conventional crops are better than organically grown ones. Far from it, my emphasis here is solely on the folly of rendering organic food inorganic and to make it clearer why cooking is so plainly an unhealthy choice. Of course, if you can source good organic food and leave it organic—i.e., eat it raw—then the more the better!

The Real Cause of Dis-Ease

Unfortunately, our bodies become increasingly addicted to the new chemicals available within these heat-modified foods, an addiction that generally lasts from the cradle to the grave, with the grave consequently being reached far quicker than it ought.

Because these foods now contain addictive substances, the body's appetite becomes warped and, consequentially, we often eat much in excess of our true needs. Unable to correctly process these biologically reconstructed foodstuffs, to break things down efficiently, the body, overburdened with these inadequately digested residues, consequently becomes latently sick.

These incorrectly absorbed foods get stored throughout the body as a residue we all know and generally loathe with a passion, called mucus.

Mucus

It matters little how a disease is classified, what Latin name it is cleverly christened, what its symptoms are, where it manifests itself, and its intensity and severity. There is one culprit whose presence at the scene of the physiological crime is generally predictable, though, perversely enough, widely ignored. It customarily resides sluggishly with few able to recognize the role it has played, and rarely gets the finger pointed at it. That miscreant is mucus.

Instead of recognizing mucus as the delinquent no-gooder it is, its presence is rarely even acknowledged. The body though, with its own innate wisdom, knows it should not be there and will do its utmost to expel it from whichever exit point is most readily accessible. As a general rule, this is done through the mouth and nose, but while mucus is sluggishly sludging its way through the bloodstream, a simple cut on the knee will suffice for the body to attempt to be rid of it.

Look at that yellow pus oozing its way out of a wound. Compare it with the pus that weeps, honey-like, from one's nose during the regular seasonal bout of colds. The difference is minimal and more one of viscosity and consistency than anything else.

Undoubtedly there'll be the occasional forlorn biologist and learned layperson who, reading this, will guffaw and protest at my apparent oversimplification, my blatant equalizing of mucus and pus, that they know through their microscopes and scientific journals to be two distinct substances. Put aside the dead leukocytes, glycoproteins, and immunoglobulins though, and one thing that can readily be shown is that on a fully raw diet, the body will not be producing anywhere near the same quantities of either.

Think of the countless fruitless millions that have been spent to combat common colds and flus—futile efforts, because the common cold will never be cured; it can only ever be prevented, since the cold itself is the cure. Prevent the cold and you prevent the onset of virtually every other medical disorder. No more cancers. No more Alzheimer's disease. No more Hodgkin's disease.

This may sound farfetched, but the truth is simple and its application and effects, profound. Every reaction of the body is a healthy reaction; your body wants to be in perfect health. Do you?

Germs

In retrospect, I am quite baffled as to why I never previously asked myself the very deep question of: "Where does all that mucus come from?" It originates from third-degree-burned food that the body has been unable to successfully process, and which has been stored throughout the body as hard and fast fat. At times this hardened fat loosens and is released as a mucus slime, becoming the ideal home and terrain for germs and bacteria to thrive within.

Contrarily, allopathic medicine would have us believe that colds are caused by invisible little airborne sprites—microscopic beings that invade our bodies, causing us untold discomfort. Not altogether antithetic to primitive thinking that the devil takes possession of you. The medical establishment would have us believe lots of things.

How fully we have accepted the explanation of imperceptible burrowing beings. But prepare yourself for something alternative, something profound. Are you ready for it? Grasp the sides of your chair. Germs can only survive if the terrain for them to thrive in is already present.

It is rather like having a dirt-encrusted dustbin that is infested with flies. From the medical industry's perspective, the flies are the nuisance and must be combated and destroyed. They prescribe fly killer, which, once sprayed inside the dustbin, will kill the flies, leaving the dustbin to rot in its own filth. After a while, other flies will arrive, or cockroaches, or more. If the bin is never cleaned, no amount of pesticides will ever see it truly clean. The super-fly will forever mutate and find its way back home!

Similarly, while mucus is ever present, things will find their home within it. Cozy, warm, and sludgy. Yippee!

In fact, the germs, like the flies, are really doing us a favor. Although irritating, they do serve to slowly but surely clean away the decay; and with each batch of germs, the body, with renewed vigor, will try its darnedest to eradicate the mucus that it breeds. The idea of simply repressing this eradication is, frankly, unreservedly ludicrous.

You would be grossly mistaken if you were to believe that this is a new idea, one that I've just concocted. It has been recognized since the days of Louis Pasteur, the guy who came up with the rather sick idea of pasteurizing stuff. Although his proposition has widely sunk into obscurity, sometime after the middle of the nineteenth century, Antoine Béchamp, a contemporary biologist of Pasteur, came up with the theory that it was the terrain that was the real issue needing to be addressed. In his own words: "Le microbe, n'est rien, le terrain, est tout." The germ is nothing, the environment (it thrives in) is everything. Pasteur and Béchamp were largely in disagreement and competition with one another over this. However, on his deathbed, Pasteur finally recanted, saying that Béchamp had been right all along, and that the terrain was indeed everything.

However, I'm guessing that since germ theory was seen as being financially far more lucrative a venture—far more profitable—the pharmaceutical industry has chosen to fully dismiss his final testimony as the insanity of a dying man. We should all suffer such insanity!

Hereditary Issues

One might contend that certain issues are hereditary and thus inescapable. One has to understand, though, that the body is like a finely tuned piece of machinery. It needs high quality oil to keep it running at optimum efficiency. Start feeding it low-grade oil, and any dirt flecks and impurities within that oil will get caught up and accumulate at the weakest spots within the machine. This is why people apparently suffer from different indispositions. On the surface it may look like the guy with the dodgy knee and the guy with the heart problem are experiencing completely separate issues, but the reality is that the sufferings of both are profoundly connected. Hereditary issues are nothing more than inherent "weak points" passed on from generation to generation. By re-adopting healthy living habits, they can quite easily be avoided.

Digestion and Assimilation

Physiologically, the anatomical body is generally occupied with one of these four activities: digestion, assimilation, elimination, and healing. Generally speaking, although all four are regularly in process somewhat simultaneously, digestion and assimilation take precedence.

I think we all understand the role of each of these two: digestion being the process by which food consumed/ingested is broken down into absorbable microscopic elements, and assimilation being the process by which those elements are then reassembled and incorporated within the body.

Because we eat the wrong things, and often in great excess, these two processes generally take up more time than they should, thus impeding the other anatomical functions where they are put on the back burner and mom-entarily neglected.

Elimination

Without the required constant flow of elimination, the body will become, for want of a better term, burdened, or increasingly more toxic. This is more than just intestinal constipation, as the internal waste needing evacuation is stored all throughout the body, and it is this gradual increase in toxicity that causes the undermining of our health for our bodies to become deformed, misshapen, puffed up, and for one's natural beauty to inevitably wane. It's sort of the opposite of that H.C. Andersen story—we all start out as beautiful swans, but after being slapped, prodded, poked, kneaded, stuffed, puffed, and corrupted, we end up as uncomely ducklings (perhaps dumplings would be more appropriate than ducklings. "Ugly duckling" is an oxymoron; these two words basically contradict each other, rather similarly to "military intelligence").

Predominantly, normal elimination involves basic usage of the intestinal and urinary tracts with the kidneys acting as filters. This process on a healthy diet is generally smooth and trouble-free and should be altogether void of any discomfort. Because of our warped sense of what constitutes food though, and because elimination is frequently postponed indefinitely, other major elimination organs are called into play.

Detox

You don't have to be a brain surgeon to understand it—just a rocket scientist (only kidding)!

Eradication of internal filth, detoxification, or simply detox by which it is commonly referred, is a sort of emergency elimination. It occurs when the body, with its own innate wisdom, decides to slow down other tasks, and instead concentrate the bulk of its efforts on elimination. It is generally accompanied by a consequential loss of appetite, such that the body can stop digestion and assimilation, and focus purely on the much-required elimination and ultimately healing (more on this soon).

Some of the more common symptoms of detox are nausea, headaches, stomach cramps, dizziness, blocked sinuses, and much more. The common cold is probably the most common form of major detox the body regularly initiates; without it, internal affairs would become increasingly more serious. The body may even become so saturated with toxins that a sudden keel-over-and-die scenario would not be out of the question eventually. It is generally unwise to attempt to suppress detox through supposed medicine or pills or any other method, and it is important to understand that it is a normal healthy bodily reaction. Your body knows well what it is doing; trust it, and don't try to fight it. Contrarily, one should rejoice that the body functions so well and that its internal wisdom demands a time-out to cleanse itself.

So, although we inevitably feel horrible, miserable, and negative throughout the detox experience, in reality, we are getting better.

Don't despair; be aware!

Although germs may be present during detox and seemingly compound the issue, they are not the enemy. In fact, they often serve to provoke one's physiology to go into a (beneficial) detox mode. Again, making a comparison to something external from biology, it is like a house full of messes and dust balls desperately in need of a good spring cleaning. It may be the case that the occupant is just too lazy to undertake the cleaning, but introduce a plague of cockroach or mice and suddenly there is increased motivation to tidy. Similarly, the germs need not even be present. The body may just decide there is too much internal filth and begin detox, regardless of the presence of pestilence. This is why you can move into a cabin out in the wilderness, far away from the city and any of its bugs and even though there are no microbes present, you can still catch a cold!

I hope my little analogies can be forgiven, because here's another one: Let us imagine, for a second, that there is a little worker whose job it is to break up twenty-five cardboard boxes an hour. The boxes are crisp and clean and it's not a tiresome task. It's an easy job and the little man is happy doing it and can do so indefinitely, day in and day out.

Let us suppose now that the little man starts being given thirty boxes an hour. Suddenly he can't keep up any more. And what's worse, these thirty boxes are no longer crisp and clean; they are dirty, soiled, hot, and sticky. The little man gets overwhelmed by it all. He tries his best to keep up, but ends up having to store decaying half-broken-down cardboard boxes all around him. In the end, he has so many encircling him that he has to shout, "Stop!" He needs a rest.

This is called a healing crisis. The little man is our body, and the cardboard boxes are what we eat. The crisp, fresh ones are the food that nature intended for us, unadulterated by heat; the soiled ones are cooked foodstuffs, and the putrefying, discarded half-broken-down ones are the mucus. Once there are too many half-broken boxes within us, we have to stop. We have to become sick; we have to clean ourselves! Then the mucus starts coming out from everywhere. We have cardboard boxes flowing out of our ears and from every available orifice. (NOTE: This is an analogy. Don't confuse cardboard boxes with cardboard hydrates—actually, you'd think it should be cardboard dehydrates?)

If you can genuinely understand why it is that the body requests detox, then you are—in theory at least—really more than halfway toward freeing yourself from regular bouts of detox. All you then need to do is act upon that knowledge!

Major Detox Points

Here are the major points to consider concerning detox:

- Detox is a good thing, never your enemy.
- Detox is never pleasant.

- Detox can take on many different and varied forms and guises.
- Detox can be sped up and intensified by not eating (see Fasting, below)!
- Future detox can be prevented by changing one's habits, but it is generally wise to not hinder present detox.
- All consumed cooked foods will eventually/sooner or later bethe cause of some detox.
- Detox can sometimes appear to be ongoing. Especially if you eat a high percentage of raw vegan foods and small amounts of cooked foods.
- Most forms of disease from which we suffer are predominantly some form of detox or other, or a result of the lack of such.
- Detox is the body's way of cleansing itself as it strives toward optimum health.
- Detox is initiated for two main reasons: because the body decides that it is in desperate need of a "spring cleaning," and because the body is invaded by an external menace (that, incidentally, can only survive and thrive where filth is already present).
- Remember that your headache is not caused by a lack of aspirin in your diet!

Cancer

As we age and continue making the same dietary mistakes, elimination is often postponed indefinitely. Decades of eating the wrong foods, and the consequent enormous backlog requiring elimination, ultimately lead to more serious complications, cancer being one of the most popular and feared afflictions.

Perversely enough, billions of dollars continue to be spent on cancer research. Sausage sizzles and cake stands are set up to raise money for the cause, with apparently zero recognition that it is precisely the sizzling sausages and the refined sugar and wheat cakes that are part and parcel of the problem erroneously trying to be cured. Yes, cancer is no more something that can ever truly be cured than is the common cold. It can only be prevented, and if the right habits are adopted, the body will rid itself of the cancer, which is precisely why fasting and healthy changes to one's lifestyle have been known to heal patients suffering from all manner of ailments. No! All we have to do is stop feeding the cancer, and it will naturally disappear of its own accord! What's that? You don't believe there is a connection between diet and cancer? Although science has yet to realize the full extent to which diet and cancer are related, it is slowly beginning to see the truth.[51]

I believe that, while the cause of cancer lies mostly unaddressed, no investigation and research into T cells will ever amount to anything of lasting purpose. Researchers may be increasing human understanding of physiology, but in terms of combating cancer, they are essentially barking up the wrong tree. In fact, the three conventional treatments for cancer—chemotherapy, radiation treatment, and surgery—have all been shown to instigate cancer when performed on an otherwise seemingly healthy body! If you are having trouble believing this, please watch GreenMedTV's informative video "Chemotherapy Doesn't Work 97% Of The Time"[52] with Dr. Glidden, who explains quite clearly that chemotherapy doesn't work in 97% of cases! With this in mind, it seems to me that such treatments are more about squeezing money out of the system than actually healing anyone.[53] In my opinion, I think diagnosis often compounds the problem, and many people would live longer not knowing they had cancer and could heed the uncommon sense I share here, rather than being diagnosed and seeking conventional treatment for it.

Chris Wark, who was diagnosed with stage three colon cancer, gives his powerful testimony of how he cures himself not through the conventional path of chemotherapy but by simply adopting a lighter vegan diet.[54]

Fasting

When we transit from the standard cooked-food diet of zombie flesh and stodgy lasagna dumpling pies (or whatever it is one eats), on to a lighter diet, whether it be vegetarian, vegan, raw food, or a fully-fledged fruitarian one, detox is pretty much inevitable. Any such transition should give the body a reprise. Easing back on the routine tasks of digestion and assimilation allows the body to instead turn its concentration toward elimination. If we desire, however, to really speed things up, fasting is a surefire way to rapidly bring on detox.

If undertaken correctly, fasting is about the closest one can get to a true "cure-all." But if you can truly fathom what's happening, then you'll know what a misnomer that is. Once one understands that the body is constantly struggling toward perfect health, and that the only thing preventing it from attaining such an ideal state is our minds and the bad habits that we have embraced and adopted, health will return of its own accord, wagging its tail behind it.

Often simply skipping a meal or two will suffice for its repercussions to begin. Digestion and assimilation will trickle slowly to a halt, leaving elimination to finally take the helm and kick in at full throttle. In fact, this is sure proof that one's body is in a toxic state. A truly normal healthy body should easily be able to skip a meal or two without any discomfort.

Fasting opens up the proverbial can of worms that the body so desperately needs to rid itself of. Once elimination is under way, discomfort will, sooner or later, be experienced. This has nothing to do with the body starving as is often supposed. It is simply a sign of the body being toxic and of those toxins being released into the bloodstream for elimination!

Fasting does have its drawbacks though, and unless you are totally sure you are up for the experience, I wouldn't recommend anyone do it without supervision. The problems with fasting are the internal demons. And I'm not referring to germs this time! For most it will, initially at least, be a constant struggle. Cravings will rear their ugly heads, and resisting them will be a genuine challenge. Once you decide to break the fast—although there is a right way to do so, easing one's way slowly back into foods, through juices first etc.—many will find the urge to binge irresistible, often undoing whatever good work had otherwise been accomplished.

For this reason, it may be easier, with milder detox symptoms, to go on a fresh fruit juice-cleansing cure instead.

Juice Feasting[55]

Whenever someone talks of doing a juice fast, or fasting on watermelon, or fasting on grapes, they are, semantically speaking, fooling themselves. The term juice fasting is another classic example of an oxymoron. Think about it: many of us may actually break-fast (the first meal of the day after a night of not eating) on a glass of grapefruit juice, or a bunch of grapes. So with that in mind, it hardly should be considered that a week on orange juice is actually fasting. You would be breaking the fast with each glass you'd take!

With this in mind, some years back, I coined the term "juice feasting" I liked it because it was just that little "e" (f(e)asting) that made the difference, and since then, I've noticed that in more recent years the term has definitely been gaining popularity. There's a vast distinction between fasting and going on a juice feast!

Fasting requires that one take no foods, solid or liquid. Only water is permitted unless one dry fasts, which is even fiercer! On a juice feast one is never fasting, but can drink as much juice as they feel comfortable with.

By making this distinction, I am not trying to undermine the benefits that can be gained from juice feasting. They are somewhat similar to fasting, only radically less intense. In theory, there is no reason why one cannot go on a juice feast indefinitely, especially if a wide range of fresh juices are consumed to cover one's nutritional needs.

I realize there is a great deal more that could be said on the subjects of both fasting and juicing, so I recommend those who are interested to read up more elsewhere. I think you might find Herbert Shelton's work on Fasting and Sunbathing quite inspiring. It can be read online for free.[56]

How Long Does Detox Take?

During my years as a fruitarian, I have been asked on numerous occasions how long one can expect detox to go on for. This is rather like the proverbial question, "How long is a piece of string?" There really is no straightforward answer.

How long it will take to get totally clean will be different for each individual, taking just weeks for some, and many months or possibly even years for others. It will depend on age and state of health and environment, the quality of food eaten, and surely other factors that are too numerous to mention. As a rule of thumb, the younger and healthier one is, the easier and smoother things should flow and the quicker the benefits should be visible.[57]

In Edmond Szekely's book The Essene Science of Fasting and the Art of Sobriety,[58] the story is told of Jesus (supposedly an Essene, thus by some definitions, borderline Fruitarian with his dietary wisdom), who took people into the wilderness to fast. There, he told them that for each year they had sinned and indulged in eating impure foods, they would have to fast for one day to recover. Thus a forty year old, to regain the given birth right of health, would need to fast for forty days.

Of course, this was just a rule of thumb, and in today's more hi-tech society—with its car exhaust fumes, chemicals in and on everything, and its excessively degenerative, denatured, greasy, fast food, burger pizza fried lard, pasta salted sugar concoctions—I have no doubt that this basic rule of "one day for each year" would no longer be remotely workable in the harsh stodgy reality of twenty-first century gastronomy.

And, of course, this was a rule of thumb while fasting, which, as I've stated, is probably the most intense form of cleansing the body can undergo, leaving the body with little choice other than to fully focus on detox, elimination, and the inevitable healing that comes with that.

So back on a juices, raw vegan diet, or fruitarian diet, how long do you expect it'll take to fully liberate oneself from previous flaws in one's diet? How long do you think it might take before those erred foods finally finish their influence upon us?

Of course, there is no single answer to this question. It would all depend on one's age, state of health, environment, and more. The answer may vary greatly from individual to individual, but I'm afraid one thing is for sure, most people quit far in advance of the visibility from the light at the end of the tunnel.

They walk away, disheartened and disillusioned. They believe that their newly chosen diet is to blame for their apparently waning health, but they don't realize that in order to succeed and once more regain an optimal state of health, all manner of physical ailments may have to be endured while the body cleanses itself.

We all need to understand the true impact that the conventional diet will have on our frail physiologies. Actually, I take that back; I think our physiologies are far from frail, given the abuse they are somehow miraculously able to withstand over the average course of a lifetime.

From my own life experience, I recall that after vagabonding my way south, from Norway to the French Pyrénées (an event that took me approximately one year and five thousand kilometres on my push-bike), I settled for a while in a French Community called "Douceur et Harmonie."[59] Upon arriving, I fully believed in, and tried my utmost to practice, the fruitarian philosophy, but was met with a good deal of opposition there.

I admit, I fully understand their skepticism, as very clearly at the time I was still regularly battling cooked food addiction, while otherwise struggling to live on a raw vegan, predominantly fruit diet. The consequences? Well, I was pretty much constantly detoxing and thus was undoubtedly far from being the great vision of health I so desperately desired to be. I also had a couple of other major health issues that the universe threw at me, and although both were completely unrelated to the fruit diet, the severity of them sent the community into panic mode, believing that my lack of culinary wisdom would be the death of me.

I reckon this was the only time in my life that I've ever experienced anything that bordered on depression. Probably due to the ever-present detox, my emotions were pretty turbulent at the time, and I often experienced anger too.

Thankfully, I eventually had enough determination to move on with my life and pursue my destiny. I see everything that has happened to me to be significant in some way, and praise be to the great white spirit—that despite any hardships that I've endured, I've managed through it all to stay focused on what I believe is right, and despite that my own journey to fruit took so many years, it is truly joyous to be here and to know that the path is forever unfolding.

So please, please don't expect to see immediate bright and shining health. A sound understanding and a great deal of self-patience will likely be required first!

Enemas and Laxatives

Patience is one of those frequently lacking virtues that will cause some to seek out ways to speed up the detox process. Colon cleanses through the likes of enemas and laxatives are two such methods frequently turned to. On the surface, such procedures may seem harmless enough, with the effects quickly made obvious; however, they should be avoided at all costs. Once you have stopped polluting your body with the lower foods, especially animal cadavers and cooked breads etc., detox will occur naturally at its own rightful pace. Speeding things up by such methods will likely result in weakening the natural workings of the body, making them lazy and dependent on them. Raw foods and plenty of juices may be slower, but even the most stubbornly encrusted mucoid plaques will slowly dissolve and eventually dislodge themselves from the walls of the intestines. Let nature take its own course!

Healing

Rather like elimination, healing, the fourth of the body's internal tasks, often gets postponed due to the ever-present digestion and assimilation demanded by a body given inferior food choices. In many ways, elimination and healing are deeply entwined with one another, and once elimination is under way, so too is healing. However, besides healing from the ravages of toxins and excess mucus, there is another kind of healing that also gets delayed regularly. I am referring to, of course, cuts, sprains, strains, twists, pulled muscles, bruises, and more.

If one is eating well and living well, healing from such wounds should be relatively quick and none too objectionable. True, one may not be able to entirely prevent pain, but it can certainly be minimized. If, on the other hand, one eats a more conventional diet, the reverse is true. Healing is abated and the level of pain experienced often increases dramatically.

There are cases of scars that, having been there for decades, suddenly heal after one has made positive improvements to one's food intake. I recall Arnold Ehret, who wrote in one of his books[60] about a friend of his who, after walking for an extended period in the mountains, breathing clean fresh air and living on predominantly grapes, experienced an almost miraculous healing of his lifelong affliction of stuttering.

I also once had a badly injured ankle, which, months later, after finally healing for the most part, left me with a slight limp. Part of me resigned myself to what seemed like the fact that I would likely never have full usage of it again. However, after some days of water fasting and the accompanying detox that had quickly reared its ugly head, jamming its gnarled clubbed foot in the door, my ankle suddenly became bathed in warmth. Half an hour later, to my surprise and pleasure, all movement had returned, and my ankle had completely healed.

What we eat not only affects us physically, but also mentally and behaviorally. This has been demonstrated quite effectively by prisons that have adopted vegan diets for inmates, resulting in profound changes of social demeanor.[61] Similarly, psychiatric hospital patients have also been shown to benefit from being fed vegetarian diets. I would love to see what could be accomplished if such institutions adopted raw vegan or, better yet, fruitarian diets!

The Medical Industry

Although I wouldn't do away with it entirely—because I do believe they do an otherwise good job in the emergency ward—putting people back together after serious injuries, stitches, realignments, and helping set fractured or broken bones, and all that sort of stuff, is about as far as my allegiance to them goes. Outside of the ER, I see the industry as a whole as little more than a glorified institution of legalized drug pushers, employed by the state to do just that: push drugs, peddling their wares on a public that worships them in an often godlike fashion. If a doctor says something's true, it unquestionably must be!

In training to be a doctor, it is often the case that very little emphasis is placed on the study of nutrition.[62] In fact, only 30% of medical schools in the US require a separate nutrition course, and on average, students receive only 23.9 contact hours of nutrition instruction during medical school.[63] It is clear from these figures that the industry as a whole directs little attention toward diet and the role it plays in health.

It seems pretty obvious to me that most issues people want resolved when they go to visit their local GPs, have their roots in bad diet. As such, a wise course of action would surely be for a doctor to examine the diet and lifestyle of a patient and prescribe changes to it. This should be clear to all concerned, just by looking at the warped body shape of the patient. Instead of being slim and agile, they are malformed and stiffly jointed, with waists, hips, and more bulging at the seams! My own experience with doctors has shown me that many of them suffer the same afflictions. I question how we can place our trust in anyone to heal us who does not understand enough about health to maintain their own.

Instead of making patients aware of their questionable dietary practices, GPs will generally shepherd you into their little offices, look at the symptoms of whatever it is that is making you feel uncomfortable, allocate some complex Latin name to it, and prescribe some kind of toxic substance to suppress those symptoms, effectively bashing them on the head. Outside of possibly other drugs, alcohol, or cigarettes, and advice to cut down on the fried and fatty foods, rarely is it the case that they will tell you to change your habits—habits that many doctors are themselves hooked on. The medication they prescribe will seriously not do anything to improve your condition long-term. More likely, it will end up hastening your journey toward an early funeral.[64]

This is the reason why it is ultimately of great benefit to patients when doctors go on strike. Statistics show that whenever doctors and the medical industry have done this, the mortality rate in the affected area has fallen dramatically.

During doctors' strikes in Bogata, Colombia, in 1976, the death rate fell by 35%. In LA, California, in 1976 the death rate dropped 18%. In Israel, 1973, the death rate dropped 50%. Only once before was there a similar drop in Israel and that was during a previous doctors strike in the beginning of the 1950s.[65] After each strike, the death rate jumped again to its normal level.[66]

They apparently have little to no real clue about the real causes of human physiological discomfort or disease, and seem to be working under the warped logic that a headache is solely due to lack of aspirin intake. In reality, although admittedly not immediately obvious, it is as if someone keeps banging their head against a wall, and then goes to the doctor to get headache pills. Bizarre, to say the least.

A few years back, we had a next-door neighbor who was morbidly obese. With the aid of crutches she could barely walk ten meters without sweating profusely, puffing and wheezing, and feeling overwhelmed with exhaustion. I paid her a visit one day and proposed that I might begin making her a fresh juice daily. This woman appears in our documentary, *Pure Fruit*.[67]

Surprisingly, she agreed to the proposal and, from that day forth, I began administering her with said juice, mostly fruit-based, but sometimes she had carrot, celery, and other vegetables too. And pretty large quantities of it—a couple of liters of fresh juice daily! In fact, I'd estimate that she was drinking more fresh juice than either Květa and I do on the days when we just take juice and nothing else.

Of course, she was still eating all the other stuff she does—the breads and meats and beefy, greasy, fried up, roasted, lamb buttocky things—which, despite her immobility, she somehow managed to extraordinarily prepare for herself in her small council flat apartment.

The juices were most definitely beneficial. Despite the fact that she still over-ate other stuff that clearly wasn't lessening her burden, she one day happily reported that she had begun more regular bowel movements and claimed to be eating less as the juice had replaced her in-between meals, or at least some of them. To an extent.

On two consecutive hospital visits, over a four-month period, she discovered on both occasions that she had lost two kilograms in weight. Okay, not a great deal as a percentage of her overall weight, but, as she had never before been seen to lose any weight at her regular six to eight week weigh ins, those four kilos were enough for her to understand that the juices were definitely a good thing in her life.

I recall carrying her juice to her one morning, and finding her sick with the flu and pounding headaches. It was her first major detox in ages and most definitely a positive sign that her body was improving. She'd just gotten back from a trip to her doctor, who had ever-so-wisely prescribed her a course of antibiotics. It's likely there are some readers who will miss out on the profundity of that!

This, together with whatever other medications she was on, and occasional trips in ambulances for heart or other obesity-related dilemmas she had inflicted upon herself over the course of many years, must have been costing her (or someone) a small fortune.

Clearly the establishment had no desire for her to get better, as she was undoubtedly one of their best customers. Had the goal been to genuinely help the poor lady, rehab would have been suggested or even enforced, where she would have been treated like others with obvious addiction problems. (Since we moved out from that particular living accommodation and moved up to our beloved tropical home, we have no idea what became of her.)

Looking back, I realize how almost comical it is that of all the "treatment" she received, the only thing that was genuinely giving her any real benefit was her daily juices, and perversely enough, other than the cost of the raw ingredients, it was the only thing she was getting for free.

The news recently featured another case of lunacy that caught my eye. An obese American woman had consulted her doctor about an ongoing case of particularly troublesome body odor. Apparently she smelled so bad that she had lost all her friends and no one would even sit next to her on public transport or in the bingo hall. How did the doctor solve this clearly embarrassing dilemma? He arranged surgery to have her sweat glands removed! The same kind of idiocy is routinely recommended to those suffering inflamed tonsils. And these people have MDs! If I were unaware of the financial incentive behind such procedures, I'd be inclined to believe that they were educated far beyond their intelligence.

Okay, okay, I may be excessively over-generalizing somewhat, and it should be clear that there are many well-meaning, knowledgeable, open-minded individuals who have made it their life choice to work within the healthcare industry. I admittedly sometimes get carried away with the negative side of the drug trade. To the undoubtedly many good, positive people who genuinely care for those who need it, I acknowledge your presence, and mean you no disrespect.

In essence though, I stand by the crux of what I am saying and believe that once one has learned to eat properly, doctors and their diagnoses of ailments, combined with the chemical addressing of symptoms, will eventually be a thing of the past.

Additional Benefits

In addition to the health improvements one can confidently expect to experience when switching to a raw vegan diet, there are other perks to a raw regime. I intend to go into these in more detail in following chapters, but first just think of that extra kitchen space you'll gain when you no longer need that bulky oven, the electricity you'll save when you no longer have to cook, and the money you'll save when you no longer need all those pots and pans and detergents to rid them of those ingrained grease, slime, and burn stains.

Grain

I look at bread as being an iconic symbol of much that is wrong with cooked foodstuffs throughout most regions of this planet. Of course, the mutilated animal corpses, fish cadavers, dairy sludge, and imprisoned birds exploited for their eggs are far worse in terms of what they are doing to ours and others' bodies, but this fact does little to lessen the crime of bread and rice and other grains.

Grain is the summit, the ever-present daily staple of shod vegetarian/vegan and zombie flesh-eating omnivores alike. It is the most unnatural of all vegan/vegetarian foods, highly environmentally destructive, and generally inedible without tedious prior preparations.

Its production is extremely labor intensive and its mono crop necessity causes widespread environmental destruction. Were you to be presented with a handful of grain plus a few peanuts and given the task of preparing your very own peanut butter sandwiches from scratch (i.e., growing your own wheat and peanuts, caring for them, weeding them, harvesting them etc.), you would undoubtedly think twice before agreeing to the challenge.

The whole idea of mono crops goes against the grain of nature. Look at a field of grass, and you might think it's just grass, but this would be a most inaccurately casual observation. In reality, the field will likely contain, at bare minimum, many tens of different types of plant species, and sometimes the variety will include hundreds. That's what nature is all about—diversity. Any attempt to forcibly eradicate that diversity, as is necessary when growing wheat or any other kind of grain, battles against nature, requiring either intense manual labor, or a potent smorgasbord of chemicals: herbicides, pesticide, and their ilk.

Not only is the production of grain slowly destroying our beautiful home planet, but also its baked consumption is one of the most common addictive foodstuffs out there and possibly the most major mucus-forming food eaten daily. Once your body is clean enough, all it takes is one or two mouthfuls of bread to recognize its quick and silent bodily phlegmification.

If you are going to eat grain, which as an Eden Fruitarian I don't endorse at all, but nevertheless recognize that at times they may be the better of choices, then consider instead of baking them first, to sprout them in soil and harvest the green shoots for juicing instead.

Salt

I do not believe the body has anything to gain from its consumption. On the contrary, I see salt intake as detrimental to longevity. As stated earlier, physiologically, the human body is well adapted for processing organic, "living" food and transforming it into the fuel the body needs in order to survive and maintain health. As it is inorganic, salt is no more a food than is dirt or rock dust. It is solely an appetite stimulator and serves no real purpose other than to promote gluttony and excessive drinking. Plants are rich in minerals that they are able to absorb from the Earth and sunlight. Minerals in plants are not identical to minerals in the Earth, sand, rock, or seawater, as they have been converted by the plant to an organic state by the miraculous powers of nature, and thus they are contrarily well suited to our bodies.

I remember as a kid always being told that we had to eat salt, that it was a necessity. When we run around working or playing, we sweat, and the sweat contains salt, and consequently such salt must be replaced. Back in the day, before discovering the benefits of eating raw, I bought into this theory, and the truth is that when I began consuming less salt, I suffered on occasion from leg cramps, laboring under the belief that these cramps were due to insufficient salt intake. I would often eat salted potato crisps in an effort to eliminate them. Indeed, it did appear as if, by doing so, the agony of cramps I suffered would lessen. However, I now see more clearly that it was likely the salt that was causing the cramps in the first place, being more of a "cold turkey" aftereffect than a genuine body cry for a life-giving necessity. In any case, anything the body expels through the skin or anywhere else, it does so because the stuff should not be in there!

Superfoods

I mentioned earlier about the raw food movement being hijacked by charlatans out to make a fast buck. "Superfoods" are one of their best money-spinners. They sell blue-green algae (basically pond scum), dried wheatgrass, maca powder, raw cacao, marine phytoplankton (basically whale fodder), and shriveled goji berries to name but a small handful of their wares. In my opinion this is nothing more than a moneymaking scam and should have absolutely nothing to do with the raw vegan movement. Anyone who takes any of these and claims to be one hundred percent raw is basically kidding themselves. Heat-processed foods are not raw regardless of temperatures used to heat them, and any food that needs first to be processed should never be considered ideal, least of all glorified with the title of "superfood!"

Raw veganism is about eating one's food as nature offers it, if nature doesn't offer it as edible, nor as easily obtainable, you can be sure nature never intended for us to eat it, nor to eat it in any great quantity.

Supplements

I know there are people who swear by them. These people insist that the Earth has been depleted of nutrients and that this is reflected in the foods it produces, which they believe to be equally deficient. I do not share this belief. I believe that the real essence of what ails us lies not in deficiency, but in toxaemia, and that pretty much all disease, no matter which cleverly coined Latin name it has been allocated, has the same basic cause. It is really quite simple; toxemia is the gradual (or more rapid) poisoning of the body through eating, inhaling, or injecting things that one basically shouldn't. Whether it be an outright poison, car exhaust fumes, heroin, petrol, cigarette smoke, alcohol, fried pizza bagel pasta pies, coffee, or even plain old mashed potatoes. All will eventually pollute the body enough to cause it disease of some form or another.

I may be wrong, but I suspect that the taking of supplements does no great favor to anyone. I'm more inclined to believe that through taking them habitually, one's constitution becomes lazy, slowly lessening the ability to absorb nutrients through natural foods—in some ways, not too dissimilar to how enemas make the bowels lazy! Perhaps as a last resort, or dire straits measure, supplements may serve some purpose, but as a general daily addition to one's diet, when eating otherwise correctly, I would not personally sanction their consumption.

I also understand that any scientific research conducted, biased or not, is based on a test subset of people who are all latently sick (they are all at varying levels of toxicity due to their flawed foods), and thus test results and statistics cannot yield truly accurate informative results.

Other Raw Regimes

Raw veganism is not the sole form of raw diet being practiced—far from it. There are people out there experimenting with all manner of raw diets.

ANOPSOLOGY[68]

Originating from France, there is a group of people practicing raw food eating, and their movement is called "Instinctothérapie." which gets translated to English as "Anopsology." The followers call themselves instinctivores, or "instinctos" for short. As well as all the stuff you'd expect to find on their tables, they habitually eat raw, sliced animals too. The flesh is not just raw, but also left to lie around fermenting for a few days before consumption. Their theory and reasoning is that primitive homo sapiens were never themselves hunters, but belonged primarily to the collection of species classified as scavengers.

Carnivorous animals basically divide into three distinct groups. First, those who hunt and kill. Second, those who hang around and wait for the killers to have their fill, so they can move in and feast on the leftovers. Lacking the correct appendages, this is where we all supposedly fit in. The last lot are the rodents and worms and maggots that eventually help to clean the carcass from sight.

Now, I'm not going to argue one way or the other about how humans may or may not have eaten in the distant past. I guess I would not be beyond persuasion to take a time machine and journey back to see us in daily action, but I'm not holding my breath waiting for the patent. And, personally, I also admit very openly to extreme skepticism concerning the whole Darwinian evolution theory.

But, even if I were convinced that we did once upon a time hang around with grumbling tummies, waiting for the wolves to finish their dinners first, it still would make no difference to how I feel today. No theory will persuade me to feast upon the flesh of others as the instinctos habitually do. No one will convince me that I must eat another or else die prematurely of malnutrition. I know that this is not the way the universe runs. Eating flesh is the grossest of all diets. The universe is built on love. Love is what makes it all go around. Even the most cold-blooded, murdersome individuals would not survive in their present corporal forms without love. Love is the driving force. Taking the life of another and feasting on its flesh is the greatest of atrocities, and is totally out of sync with our true essence, which is nothing short of Divine hug-filled loveliness.

To give them credit, the instinctos are generally of above average health, but this does not in the least surprise me, as the average is far lower than it could or should be. And to give even more credit, the main principle on how they eat has been a major inspiration for me and helped refine my diet to its current transitory form.

Generally speaking, they let their noses have first choice. Whatever on their raw tables smells most appealing is given highest priority to the mouth. They will eat a particular food, whether it be oranges, carrots, durians, dead sentient being flesh, birds' eggs, or plates of wriggling insects, and they will continue feeding on that one particular foodstuff until the body, through subtle messages, indicates that it has eaten enough. This may be signaled by a slight tingling on the lips or tongue, a change in taste, or another message from deeper within the body. This they call l'arrêt instinctif.

The Instinctive Stop

Once the instinctive stop has been reached, they will push their food aside and, if they still feel their appetites have not been satiated, they will return to their tables, once more smell each item, and choose the one with the most appealing odor. The pattern will continue, such that one meal may consist of two or three different types of food, until their hungers are satiated. Although I can't say that this is how I too choose my meals, I can see the underlying sense in doing so.

Instinctive eating only works provided your choice is of foods in their natural states. As said, the instinctos eat wriggling maggots, live insects, rotting fleshful morsels, fish in decomposition, and birds' eggs, and consider it important that these be included regularly whenever the nose doth choose. Their philosophy accepts neither milk nor any dairy stuff to be valid table options, as these were clearly, to them, never a part of our original diet. It is a very well structured method of eating, but lacks compassion. It is purely based on one man's thoughts (Guy-Claude Burger[69]) on how he believed primitive man once ate and lived (their philosophy covers more than diet alone).

It is also a very social method of eating, as it cannot generally be practiced fully by lone individuals since very few of us have the economic wealth enough to load our tables with broad ranges of foodstuffs for each meal.

Instinctive eating can only be practiced with a valid selection of foodstuffs. Given also the choice of mixed conglomerations, combination abominations, and cooked monstrosities, the nose will generally lie to one's body about what it needs. These types of smells were generally never around at the dawn of time, so the human body does not know how to analyze correctly their value as food.

Our sense of smell is a highly powerful organ rarely given its full credit, and my overall feeling is that there is much to be learnt from the instinctos, but that they themselves have a lot to learn about the true nature of life.

Paleo / Primal

Similarly to instinctivorism, the Paleo Diet[70] and Primal Blueprint[71] are based on so-called evolutionary science—the theory of how primitive humans once ate—and attempt to guide followers to adopt the patterns of pre-agricultural ancestors. The two diets are often confused, as they have many aspects in common, but the basic difference is that the Primal Blueprint diet allows dairy, whereas Paleo doesn't. I mention them both in passing, but as neither have a true understanding of the Golden Rule (explained in more detail in the chapter on religion), similar to the shod omnivores, they regard their food purely as food and disregard any sentience or objection that food might have in being eaten. I will not spend more time on them.

801010

The 801010ers[72] (pronounced eighty, ten, ten-ers) are a more recent group of almost raw dietary vegans. The movement was founded by Doug Graham who also authored the book The 80/10/10 Diet.[73] Many of the 801010 crowd falsely refer to themselves as fruitarians. I intend to go into more detail with this in the next chapter.

Although not all of those claiming to adhere to this regime are themselves vegan, 801010 is also referred to as a LFRV diet (Low Fat Raw Vegan). The meaning of the numbers is a basic calorie guideline, where it is recommended that 80% of one's calories come from carbohydrates, 10% from protein, and the final 10% from fat, and that those calories should come from raw fruit, vegetables, nuts, and seeds.

There is no doubt in my mind that this conception has been of great assistance to many, encouraging and helping them to improve both their diets and lifestyles; however, for my own personal taste, I find that as well as grossly lacking in simplicity, (despite frequent claims made to the contrary by core members of the group), its focus predominantly concerns health and fitness, thus it lacks the crucial empathy component of veganism itself.

As I've stated in my introduction, I have no formal education, and to me all this talk of calories is just meaningless hot air. As a fruitarian, I also have no desire to even look at my food as a bunch of calories or carbohydrates, proteins, and fats. I wouldn't recognize a protein under a microscope, and carbohydrates don't sound like anything I'm seriously ever going to need. Referring to fruit in such terms removes their beauty—effectively semantically sullying and insulting them.

But hey, that's just me, and I guess whatever floats your boat or turns you on.

They often talk of fruit as fats and sugars too, which I also find degrading, and although I've always been somewhat mathematically-minded, I still find all that counting that they do confusing. I think they have a sort of minimum guideline for the amount of calories one should eat daily—three thousand—and are always talking of "carbing up" and of high carbs. They most definitely encourage overeating, recommending that when you are full, you force yourself to eat just a tad more, just to be sure, and consequently discourage one to trust in the body's own innate wisdom to choose what it needs given a valid selection of foods.

I state that they are "almost" raw, as some include cooked rice, and most if not all include nuts, seeds, and dried fruit (more on this soon); and I make the distinction that they are dietary vegan, as the 801010 focus is predominantly on fitness and health, and thus there are clearly no guidelines discouraging leather belts, trips to zoos and circuses with animals, keeping pets or vivisectionally enhanced cosmetics, or even using non-vegan B12 supplements.

That brings us to the next flavor of raw thinking and the crux of this book: Eden Fruitarianism.

I don't believe in killing whatever the reason!

- John Lennon

Chapter 5
Eden Fruitarianism

I began the previous chapter by telling you how raw veganism can also very much be an ethical choice. That is, besides the very valid moral duty we have to keep our own bodies healthy. Both raw veganism and fruitarianism are missing crucial ingredients if they do not also focus on widening one's powers of empathy, ergo compassion and kindness. I intend to lemon and limelight these issues throughout this chapter.

I believe it of utmost importance that the moral guidelines and health aspects outlined in previous chapters be fully understood before one can be ready to succeed with Eden Fruitarianism, so if you have impatiently skipped forward to this chapter, I can tell you now that you are missing out on the solid foundations around which this philosophy is constructed. If you haven't already done so, please take the time to read the previous chapters before continuing.

Fruitarianism

Firstly, there are probably as many flavors of fruitarians as there are of vegetarians. We get vegetarians who eat everything but red meat, and others who eat chicken and fish, and of course the vast majority who eat dairy products and eggs. And yet the word itself sounds like it would depict someone who just ate from the vegetable kingdom. Far from it. Language is often illogical.

Similarly, we get fruitarians who eat roots, leaves, tubers, seeds, and grains, indeed, even whole plants. Shucks, there are people calling themselves fruitarians who cook or otherwise abuse their foods before ingesting, and even those who are not totally opposed to eating the occasional piece of flesh!

As a little anecdote, I recall some years ago that a couple of friends of mine went to meet a self-proclaimed fruitarian living in Ireland. They found her at home eating bread with peanut butter and jam. In her mind, her meal was made entirely from fruit and seeds/grains that she supposed to be a regular part of a fruitarian diet!

There are fruitarians who mix it all together in astronomical gastro-nomical onslaughts, and others who try to simplify by eating mono, and some who call themselves fruitarian that will even tell us that it is dangerous to eat just fruit!

I suppose it is just the way of the world, and that all teachings are eventually perverted, distorted, warped, and diversified. I feel quite sure (admittedly, perhaps naïvely so) that original vegetarians just partook of vegetable foods and never dreamed of supplementing their diets with milk from other species, or birds' eggs. Certainly they would never have considered themselves vegetarians if they chewed and swallowed fishes or birds.

But language adapts itself, and people who weren't vegetarian became vegetarian, not by changing their diets but just by the constant redefining of the word "vegetarian," until they became socially and linguistically accepted as being vegetarians, even though they clearly once wouldn't have been.

Thus, out of necessity, new words spring into existence to help define where one really, diet-wise and ethically speaking, stands. The word vegan is a good example, popping into existence in the 1940s, to describe someone who was really vegetarian, and later still the terms lacto-vegetarian, ovo-vegetarian, pesco-vegetarian, lactov-ovo vegetarian, and even semi-vegetarian and socio-vegetarian became somewhat common descriptive tags.

I fear that much the same has already happened to the term fruitarian, with people claiming that if 80% of one's diet is fruit, then one is fruitarian. This random percentage has been played with extensively to allow more and more people to consider themselves fruitarians, such as 75%, 70%, 60% fruit. Make up your own percentage and become a fruitarian! Then I've seen people calling themselves lacto-fruitarians, which a couple of Swedish contacts I had in the past believed to be a sound version of fruitarianism, which involves capturing, imprisoning, and stealing the milk of other species. Of course there are all the 801010 fruitarians too, notorious for telling people that 100% raw fresh fruit is not only infeasible, but also dangerous, and many still haven't fully grasped veganism, let alone raw veganism.

Looking at the word fruitarian, you would think that it means someone who just eats fruit. Agreed, that seems logical enough, but the truth is that the very vast majority of people out there who call themselves fruitarians, do make exceptions to the "just fruit" rule, most of them on a regular daily basis.

With this in mind, I'm going to be brave and coin a new term, Eden Fruitarianism.

Defining Eden Fruitarianism

Eden Fruitarianism is, I believe, the original fruitarianism as practiced by the once-upon-a-time couple, Adam and Eve. It can only be reached through a full understanding of the aforementioned "isms" (so get back there and read them if you haven't already!) and is predominantly a philosophy based on ethics, requiring the very deep realization and understanding that of all the food available in this world, the only nutritional substance that is truly given, and thus has the potential to be totally free of any violence, is FRUIT!

Plants make use of a variety of ingenious methods to disperse their seed.[74] These methods include taking advantage of gravity, wind, the flow of water, and the movement of animals (where the seed hooks itself onto the fur of a passing animal). Some plants are even able to make use of ballistic/mechanical dispersal, where a seed pod explodes, outwardly flinging its seeds away from the mother plant.[75] Probably the most successful technique many plants employ is to encapsulate their seeds in an often vibrantly colored, fragrantly odorous, deliciously tasting flesh that many animals, including humans, naturally find appealing. This is one of nature's ways to give the seed a chance to be moved some distance away from the mother plant.

Clearly, in such cases, the plant wants other life forms to consume this fruit flesh.

Bacon would not be a choice if the pig had any say in the matter. A lamb, given the gift of speech, would most probably say "no" if you asked if you could eat its leg. Fish would no doubt choose to stay in the water if they could, and I feel pretty sure turkeys would object once their Christmas fête (or should that be fate?) is made clear to them. Chickens would surely be protesting having their eggs systematically stolen and freedoms restricted, and both cows and their calves would be up in arms, if they had any, over the theft of their milk. Given the chance, bees would attack and defend ferociously, even sacrificing themselves in the process, in order to protect their precious honey—a sure sign they do not part with it voluntarily.

But even for the things that have no oral or physical ways to defend themselves, it does not take a great leap in thought and faith to understand that they too, given the ability to do so, would likely protest.

Eating a carrot involves destruction of the whole plant. The lives of lettuces and cabbages are brought short once picked. Even individual leaves torn off a plant seem, in some way to me, to be against the grain of a universe I believe to be built on love. Eating grains, cereal, nuts, and seeds generally involves depriving those capsules of any real chance of a life that could have been theirs.

Plant Perception

If you believe that such plant life, lacking an obvious animal comparable nervous system, cannot feel pain and you are unable thus to feel empathy for such life forms, then I might suggest you read *The Secret Life of Plants*.[76] This book will offer insight into generally unseen plant life communication and emotional expression.

Published in the first half of the 1970s, the book details experiments conducted on plants that were hooked up to a polygraph, a basic lie detector of the era. The results proved beyond reasonable doubt that plants would react and respond to violence toward themselves as well as surrounding plants and animals. Indeed, not only do they react to such negative stimuli, but even to the mere threat of such action! Experiments showed that it was enough to even think about harming a plant for it to react and scream out in protest. Screams that rarely get heard and recognized. If this seems a little farfetched to you, then please take the time to watch this short video[77] from the *MythBusters*[78] team who basically were able to confirm the findings of the 1970s scientists.

If, after watching it, you are psyched to see more, try this short (under three minutes) YouTube video.[79] Or, if you feel up to it, watch the full length YouTube documentary[80] (just over one and a half hours) about the secret life of plants.

Also, more recently, the University of Aberdeen conducted a study[81] that showed how plants communicate amongst themselves, warning each other of approaching pests and threats.

Fascinating reading.

By showing this, I am not trying to say that taking the life of a plant equates to taking the life of an animal. It should be clear that even though a fruitarian acknowledges the consciousness of a plant, and attempts whenever possible to minimize harm to any, we are likely all able to weed or end the life of unwanted plants, despite that recognition. Faced with the choice of taking a knife to the throat of a calf, or pulling up a carrot, no fruitarian will wonder which option is worse. Of course, if fruit is available, then such choices need not be faced at all.

The Uniqueness of Fruit

Move on to fruit and you'll notice that it is a whole different kettle of fish. Or should I say, "basket of mangoes!"

Going once more back to the 1970s—toward the end of the decade this time, back in the days when I used to work on giant computers in large false-floored, air-conditioned rooms—I got hooked on a computer game. It was probably the very first Dungeons and Dragons game ever written. It was a single user game called Adventure, or Advent for short. The game involved a labyrinth of rooms and corridors with a surprise around each corner. One of the creatures regularly encountered in that maze of twisty little passages was a troll. The trolls were unique in that the only food they ate was coal. Back then, I used to think that all food was taken against the best wishes of whatever life it came from, and that the world we lived in was unavoidably cruel. The trolls, I thought, had evolved something wonderful. They just ate coal and, in theory, were able to live totally at peace with their surroundings.

Of course, eating coal was never something I seriously aspired to do, but I couldn't help but wonder how the world might be if all we needed to eat were raw inorganic materials that were in abundance everywhere.

It took me years to recognize fruit for its comparable potential. Contrarily to other foods, given the gift of a set of vocal cords, the plant that bears the fruit would most likely be singing, "Eat them! Eat them! Here they are!" Because through eating the fruit, you will be helping the tree spread its kernels. And the tastier the fruit, the more likely you will be to nurture and care for its seed.

To my knowledge, short of becoming brethren, fruit is the only viable, usable solution to the live and let die scenario we currently find ourselves ensnared within. I repeat myself: fruit is the only food that is genuinely given, thus karmically free, and thus the only food (discounting inorganic materials, such as coal, of course) that is truly compatible with a universe whose fundamental principal and driving force is one of love.

With this in mind, Eden Fruitarianism involves eating or genuinely aspiring to eat a diet consisting of 100% raw fresh fruit. I probably still need to clarify the term fruit though, as there is clearly confusion among many as to exactly what a fruit is. I have, for instance, seen reports online of people saying things like, "I eat 100% fruit," and then, going on to more closely describe their diet, will list that they eat greens, nuts, and seeds too.

Defining Fruit

When I talk of fruit, I am referring to the often-edible part of the plant that surrounds the seed of the plant. Durians, mangos, nectarines, chocolate pudding fruits, apples, oranges, strawberries, bananas, pineapples, cherries, and other such obvious fruits are just a few of the countless species of fruits that spring immediately to mind.

Despite not always being classified as fruit, the following also belong to that category: avocado, tomato, cucumber, zucchini, peppers, pumpkin, and even eggplant and belladonna (deadly night shade). Clearly, the flesh surrounding the seed is not always edible by humans, thus there are many types of fruit that are not suitable food for us bipeds.

In some cases, such as the banana and the pineapple, the seed is often no longer present; these are, however, still types of fruit. Although rare in the Western world, varieties of both still exist containing seed.

In other cases, such as the tomato, the kiwi, the pomegranate, the passion fruit, and the strawberry, to name but a few, the seed and the edible flesh are basically inseparable, thus eating the seed too is unavoidable. However, in such cases, the seed generally stands to profit from the experience, as the digestive process will help kick start whatever enzymes are in the seed needed to get it germinating. Thus the deal is symbiotic for both us and the plant. Provided, of course, that one does not just flush one's bodily wastes into the sewage system!

What's Not Fruit?

Well, obviously not meat, fish, and eggs, nor cheesy, yogurt-y slime. Not roots, nor leaves, nor tubers, nor nuts, nor legumes, nor grains. Even if raw (and even though some do consider these to be regular parts of the fruitarian diet), they ultimately do not form part of an Eden Fruitarian's fare, who sees beyond things as being merely food or not and empathizes with the history of their meal.

Difference Between Fruit and Nuts

I keep seeing people writing that botanically speaking, fruit, nuts, and seeds are the same thing. I can sort of understand where they are coming from—by some definitions, everything that comes from the flower is considered, in some way, to be the fruit of a plant. However, from an Eden Fruitarian's perspective, although they are clearly in physical proximity, they are no more similar than a root and a leaf. Playing with words similarly, a calf could equally be said to be the fruit of a cow, and one's baby the fruit of one's womb, or even seafood the "fruit of the sea." Clearly (too many) atrocities are committed when any of these are eaten.

Think of a nectarine. What we eat is the precious, sweet nectar flesh that surrounds the stone/seed of the plant. The seed is clearly separate from the flesh. The flesh is the thing that is, so to say, given karmically freely.

Think of a melon. What we eat is the delicious flesh of the melon, not the seeds in the center that generally get left out of the digestive experience. (Yes, I'm aware that some cultures roast them, but fresh out of the fruit they are none too appealing in my eyes). Think of a papaya. We eat the life-giving, juicy, vibrantly colorful, sweet flesh of the papaya. The small black seeds get released back into nature. Or should.

Think of an apple. The flesh is savored, and the core is discarded. I could continue ad infinitum.

Clearly not every seed is surrounded by this so clearly precious substance called fruit. Some seeds/nuts/grains etc. are void of the encapsulating fruit flesh. That's just the way the world is. Not every plant consists of a fruit. Not every plant consists of a seed either, for that matter. Nature is abundant with variations.

So What's Wrong with Nuts?

Well, that's a loaded question. I'm not trying to say that anything is wrong with nuts, but then again, I'm not saying anything is wrong with seeds, grains, shoots, leaves, roots, tubers, insects, fish, mammals, or other animals. Although there might be.

However, I just don't believe any of the above are ideal foods for humans.

But about nuts specifically. Besides clearly not being fruit, I think it's well worth acknowledging that, unlike fruit, commercially purchasable raw nuts are simply not available! I've had people write to me and say, "But David Wolfe (raw food guru) is selling raw almonds or cashews!" etc. Don't be fooled. It's simply not true. If it is, then it's because he means something else by the word "raw" than I do right now.

When I say raw, I'm talking of unprocessed raw. I'm talking about raw, ripe, and fresh. When others say they are selling raw nuts, they are most likely selling processed (sun or air) dried nuts, about as raw as their shriveled goji berries.

As described in the previous chapter, all-natural, fresh, ripe, and ready foods, eaten mono on an empty stomach, have what the French have christened as an "instinctive stop." Dehydrated foods, including nuts, have no instinctive stop. At best, it's physiologically a second-class food. Your body will crave the addictiveness of them, and will thus be unable to truthfully tell you that enough is enough.

As an experiment, I'd like you to go out in nature and find some truly fresh (i.e., not even sun-dried!) nuts and sit and eat them. I think you will have had enough of them far sooner than you ever would with those picked, de-shelled, and packaged store-bought ones.

Truly fresh nuts are slightly wet and soft. Not dried and crispy/crunchy!

Thus an Eden Fruitarian avoids such dehydrated foods, favoring also grapes over raisins, plums over prunes, and fresh mango over dried, withered, dehydrated mango. To summarize, I avoid nuts for the following reasons:

1. Ethically, I'd like to give them the chance to fulfill their potential treeful destiny.
2. They are not commercially available raw.
3. I don't consider them to be physiologically ideal foods. Too heavy.

What About Coconuts?

Well, yes, granted, unlike other nuts, they are definitely readily available, raw, ripe, and fresh, and in abundance for some. Physiologically, I'd reckon for that reason that they'd have to be better for you than most any other nut available. However, they are still not fruit!

I love to see coconuts fall down when they are ready and sprout in the sand. Coconut palms are a beautiful sight. And I have personally made the choice to try and avoid ending the life of another prematurely. Of course, that's my choice and I respect that others may not feel the same way. We all have to reach our own conclusions about such matters.

I understand when people tell me there are an abundance of nuts and that by eating some we may also help propagate others. Yes, if that's how you feel about them then great; enjoy them! My belief, which I'm not claiming is infallible, is that fruit offers superior nutritional benefits and is ultimately the least harmful choice. Thus I'm choosing to stick to just fruit.

I think another thing we might observe about nuts is how difficult they are to get into. Think about it: every fruit, when it's ready, will grant you easy access to the flesh. Try and get into a coconut with your bare hands, or a macadamia nut with your teeth, and that's a whole different story. Without a tool of some kind, you stand little chance of getting at the edible interior.

In addition, although the husks may make good mulching material, the shells of coconuts take significant time to break down, and if not disposed of sensibly, all it takes is a quick downpour for them to become perfect breeding ground for mosquitoes to thrive in!

What About Legumes, Peas, Beans, etc.?

I see. Well, you can certainly get them fresh and ripe to eat, that's for sure. But I still don't consider them to be fruit, botanically or otherwise. They are seeds and, to be honest, although I've made occasional exceptions in the past, I'd rather let seeds be.

I'm choosing to be fruitarian because I believe it is the garden of Eden diet and because I'm a hopeless romantic who believes it's love that ultimately makes the world go 'round. I firmly believe that "love is all you need" and that fruit is the only food that fully has the potential to be compatible with that notion, being not only the highest of all foods (physically and spiritually), but also the only food that is truly symbiotically shared by any living thing.

There are, however, a couple of exceptions with legumes that make certain varieties suitable fare for the Eden Fruitarian. Carob and ice cream bean (inga bean) are two such exceptions, where either the pod itself—or in the case of ice-cream bean, the fluffy packaging encompassing the seed—is both desirable and edible, and serves the same purpose as the flesh surrounding other seeds.

It encourages the eater to carry away the seed from the mother plant and distribute it further afield.

Mushrooms

Suffice to say mushrooms are not fruit and thus are not the idealist of foods; however, as foods go, they are far from being the worst choice one can make. I have little really to say about them, but I think for all such similar questions, it might be wise to refer to the section coming shortly called, "The Ethical Food Tower."

Nutritional Science

Off the Cuff

I'm allergic to calories. They make me sweat and fart and froth and burp. Carbohydrates are killing me slowly, and when I look at photos from thirty years ago, I can see how much they've aged me. Vitamins have played their role too, with some hair loss, not to mention the minerals that have destroyed my teeth. And then there are those dreaded glycogens, proteins, and polypeptides. Who knows what havoc they've inflicted upon my body over the years!

Oh how I wish I had grown up in a world full of fruit and guardians with sense enough to understand its life-giving properties. I could have avoided all those ghastly amino acids and sugars that have (rav)aged my body.

My simple understanding of a calorie is that it's a unit equal to the amount of heat required to raise the temperature of one gram of water by one degree at one atmosphere pressure, and as such I fail to see what business they have playing vital roles in our diets. The mere thought of them sets my spine a shivering. It's rather like eating empty hot air, don't you think?

In my opinion, all that pseudo-science of microscopically dissecting the molecules of our food before eating them distances us from the pure simplicity of eating the fresh, juicy, ambrosial, heavenly scrumptiousness of fruit. It degrades the sweet and saporous, dribbilicious nectarine to a mound of health-threatening sugars and other pathogens. The durian gets transformed from a delectable hunk of tantalizing custard, creamy, peerless, spiritedly irreproachable bliss into a congress of lifeless decaying vitamins, ugly nefarious fats (God forbid!), and gremlinous amino acids.

Thus we have the meat and dairy industry that cleverly marketed us the protein myth—the raw fooders who sell us their goji berries with all their magical built-in ingredients, and now the new set of fruitarians who insist that we need to understand microscopic food content before we can hope to do it properly. They want for us to believe that we cannot do it ourselves, but need expert opinions and guidance to get there.

Take a simple apple and cut it in half. After barely ten minutes what do you observe? The two halves exposed to the air are likely already beginning to discolor, slowly becoming increasingly browner. Put very simply (because, let's face it, I ain't no scientist!), what's happening is the life force of the food is slowly dissipating. The longer you leave it, the less value it will have as a food.

Now, take another apple and grate it with a grater. You'll observe that the smaller the pieces, the quicker they'll discolor (oxidize) and the quicker the true value of the food is lost. If you start doing stuff like subjecting the morsels to heat or similar, then the process is yet more rapid and far more devastating! In order to observe those pesky, impalpable little vitamins and their ilk, food has to be dissected so small that by the time it's sandwiched between the glass under the microscope, you can be sure that any vital, once-present life force is long gone.

Same goes for the amino acids, glycogens, carbs, and the ever-elusive calorie? I mean, yeah sure, you might textbook quote me any of it, but admit it, really you're just "blah-ing" stuff that you have no idea about. How many of us would even recognize a plate full of pure B12 if we saw it? And yet so many tend to talk about them, feigning a familiarity that is nothing short of fictitious.

I don't care for any of it. I also don't believe that anyone can rightly say what anyone else should specifically eat at any given meal. If you think that you need someone to guide or tell you, then you surely haven't understood the simplicity of true fruitarianism. I don't care if someone tells me I ate just forty calories yesterday, or four thousand; to me it means nothing, so I'm not into calorie restriction, nor the opposite of force feeding myself, nor castigating anyone else into believing they need to get a certain amount each day. I don't give a damn how many grams of food I ate in my last meal, and I'm not into ritually forcing drink down my gullet to stay hydrated in the future.

The next time someone comes up to you and tries to get you to eat protein or calories, be strong, say, "No, thank you," and turn and walk the other way. Resist the temptation, eat fruit, live long (haired), and prosper.

Seriously Though

Okay, tongue out of cheek, it hasn't always been the case that I've avoided nutritional science. Way back, when I first became vegetarian, I was slightly obsessed with the subject, reading up on it through whatever books were out there at the time, determined to succeed with vegetarianism, followed very soon thereafter by veganism and believing that only through a sound knowledge of science would that be possible. Once upon a time, for example, I knew the name of every single essential amino acid that the human body is said to require—and not only their complex individual names but also their general function and in what vegan foods they could generally be sourced. Nowadays, I can't even recall if there are supposed to be eight, nine, or ten essential amino acids, and I don't remember the name or particular function of any one of them.

At some point along my path, I reached an epiphany: if every major change I had made to my diet was indeed an improvement, which I solemnly professed to believing, then why was I suddenly obsessed with the possibility of nutritionally failing? Surely if each step was truly an improvement, then I should have been less and less concerned about each individual microscopic element of nutrition. After all, most people go through their lifetimes without ever thinking about Vitamin A or iron, and there I was fretting over whether or not I was getting adequate amounts of both, even though, I profess, I had no real idea what either of them were. It was like a part of me didn't have faith in the path I'd taken, most likely because the vast bulk of the rest of society was in opposition to what I was feeling, believing, and practicing; and because, at the time, I lacked confidence facing that opposition.

The more I delved into nutritional science though, the more I noticed that there were an awful lot of conflicting ideas being peddled. This basically meant that, at the end of the day, it was up to me as an individual to decide which of the many and varied, opposing viewpoints I believed in, with no way to certify that any were genuinely correct, even though all were somehow purportedly based on supposed proven, unshakable scientific fact. Thus, not having any real idea what a complex carbohydrate was, nor how to verify if what I was reading about them was correct, at the end of the day it was still all about belief and trust that someone else genuinely knew what they were talking about.

It became clear to me too that whenever any nutritional science research is funded and undertaken, it is pretty much always done with some kind of ulterior motive. Probably the bulk of money spent on nutritional research has been funded as part of some kind of marketing ploy, and as such, results are sometimes achieved through methods that one might describe as mendacious, or more charitably, less than honorable. Research says that we need protein for healthy cell growth and tissue maintenance, and meat contains protein, ipso facto we encourage you to eat meat. No one seems to question how the beef got so full of protein eating just grass!

We need calcium for healthy bones and teeth, and milk is an excellent source of calcium. In today's world it's not always easy to get as much iron as we need, therefore we propose you buy our unique brand of iron tablets. See how this kind of research can and has been, on countless occasions, used in advertising? Just because it's written down somewhere, just because it's a shared common belief, doesn't make it fact! And even if someone has purportedly scientifically proven something, this does not mean that YOU know it to be true too. The most you can say is that you trust that everyone else knows what they are talking about.

Protein

In fact, we have all been constantly fed such biased conclusions that we often go around quoting them as gospel. This is the main reason people are always asking vegans where they get their protein. The question itself implies that the questioner really has an understanding of protein, something that is very rarely even remotely true. If I picture chewing on one, it is bland, dry, and like chewing an old boot. I don't know what they look like, nor do I care. Sure, I understand them to be purportedly made up of amino acids, and before the body can assimilate them, they must first be broken down once more into those self-same acids. But one has to see how distant a protein is from the reality of food. If you are walking through a garden and find a cherry tree, sit under it and enjoy the small, purple, bliss bombs until satiated, then lie down to rest on the grass, smelling the sweet wafting fragrance of flowers and listening to the melodious songs of birds, the last thing on your mind will be, "I wonder if I got my full complement of proteins."

It appears to be the case that not getting enough protein is a lot more difficult than one might expect, and to succeed in protein deficiency almost requires starving oneself. It is said, for example, that roughly 0.9% of human mothers' milk is protein,[82] this being nature's intended sole food for a human at probably the most rapid growth stage of life!

This would seem to imply that such small percentages of protein should generally be easily adequate throughout other stages of one's life too. As it happens, the protein content of a broad range of fruit is said to vary from between roughly 0.5% to 3%, which should be far more than adequate to meet one's supposed needs.

Not yet convinced? You might be surprised to know there are even raw food bodybuilders out there who manage to bulk up despite obvious meat deficiencies.[83] I read this morning that a group of fifteen vegans competed against hundreds of bodybuilders with traditional omnivorous, heavy meat, dairy, and egg-based diets. The results? The vegans won five out of seven divisions that they were in.[84]

But it isn't my intention to delve further into such questions. If it interests you, you might check out The China Study,[85] research that I'm told proves the superiority of a vegan diet! I also recommend Ross Horne's book The Health Revolution,[86] which delves more deeply into the science behind a fruit-centered diet. I am quite certain that if examined truthfully and without bias, science has the potential to one day prove beyond a shred of doubt that a 100% raw, fresh, ripe fruit diet is superior to all other diets; but even should such a theoretical day come to pass, it is likely that some will choose to not believe its credibility. Instead they'll choose to keep their faith where their ingrained habits already are. Again, in the end it is all up to us as individuals to decide what we believe to be true and false.

Even if we were to self-experiment with every conceivable diet, it can sometimes take years for consequential health problems to become apparent. Get a well-read vegan and an equally well-read Paleo guy in the same room, and both will swear blind that research has proven undoubtedly that their own particular individual diet is nutritionally superior. From a purely nutritional outlook, they could argue ad infinitum and get nowhere.

Faith

That's why I don't desire to go there. Eden Fruitarianism is about neither nutrition nor any accompanying set of pseudo-science "facts." It is about ethics and ultimately faith that there are ethically correct alternatives that are to the benefit of all sentience involved. Thus, whether we are willing to accept this or not, any nutritional choice relies on faith of some kind. Most people are not, and they mistakenly believe that they have knowledge and indefatigable science to back them up, but all they really have is faith that their information source is reliable and that results have been interpreted correctly.

I am personally choosing to forsake nutritional science entirely and place my faith instead with the concept of live and let live, which I believe to be a valid path. This does not mean that you too should do so. By all means, do whatever research you feel is necessary. Other than vague warnings that an all-fruit diet may be lacking in something, there seems to be nothing more specific than that. Not enough minerals, some will say, but push them to find out what specific minerals, and no one can tell you. Mostly they are all quoting each other, and truly valid non-biased research has yet to be properly undertaken.

To me, it makes a kind of intrinsic sense that whatever is truly best for the environment will also be best for us as individuals. Whatever is best for the ecology will be best for the long-term health of whichever economy is employed. Whatever is best for our health must also be the best for the health of the planet as a whole and of all sentience involved. The path of do-as-little-harm-as-possible will not only be the most beneficial for all those we come in contact with but also for those following that path.

I believe that the currently popularly shared belief of live and let die— that in order to live, something must die— is inherently flawed and serves no real use other than to help keep the world ensnared in its dark ages. I believe that Eden Fruitarianism offers a long-term solution to this dilemma; it nurtures empathy, and although I cannot truthfully fully, clearly envision where widespread adoption of the philosophy will lead, I have every faith that it is into the light!

This "faith" in doing the right thing for all parties present and concerned is of utmost importance in succeeding with Eden Fruitarianism. Without it, one cannot truthfully claim to be on its path. One must have an understanding that surrendering to this path will likely cause an onslaught of opposition, and one must be prepared to face that opposition calmly, peacefully, and with full faith that whatever obstacles are strewn in one's path— however thick and irksome they may seem— there is always a route through them, however troublesome the obstruction. It will always ultimately be surmountable. And remember, the great white spirit never gives us more than we are capable of handling!

Such obstacles will come in the form of seemingly overwhelming opposition from apparently well-meaning kith and kin, your own nagging doubts about whether Eden Fruitarianism is anything more than the crazy thoughts of an insane minority, or an onslaught of health issues all connected to the fact that your body is finally cleaning itself. It is understandable that without an underlying strong faith that what you are doing is of undeniable benefit, failure will be predictable. Remember at this stage that you are a pioneer; not everyone is pioneer material.

Naturally I have my own experiences of that which I talk— the onslaught of opposition from friends, the ridicule and accusations of extremism, the temptations and struggles and many stumbles, the accompanying ongoing detox, etc. All, or any of this, is enough for many to forsake the path and return to the familiarity, comfort, and security of the sheep flock. I think the occasional stumble is inevitable and I don't see stumbling, in and of itself, as a problem. The important thing for Eden Fruitarianism is keeping the faith. Lack of faith and the accompanying fear that one is failing have already deterred many. Such challenges await us all.

For an example, a few years back, while on a 100% fruit diet, I went through a period of experiencing extreme fatigue. It was literally the case that I could barely walk a hundred meters without feeling exhausted and having to sit down or rest. This went on for an uncomfortably long period of time. I could have consulted a doctor, who would no doubt have informed me I was suffering from low blood pressure and prescribed drugs to raise it or (if they had even bothered to enquire) drawn the conclusion that my diet was surely to blame, but I didn't. It's likely that low blood pressure was indeed a symptom, but as I could see no reason why I should be suffering from such, other than detox, I decided to wait it out. Sure enough, about one year later my energy levels returned to normal, without having done anything but persist in my pursuance of Eden Fruitarianism!

I also understand that any scientific research conducted, biased or not, is based on a test subset of people who are all latently sick (they are all at varying levels of toxicity due to their flawed foods), thus test results and statistics cannot yield truly healthy and informative results.

The way I see it, there is one major cause to what ails us, with many varying symptoms. The only real lasting solution: stop the pollution!

I think humans have barely scraped the surface in regards to our limited knowledge of how the body functions and all it requires to do so. For an example, I used to know this guy who had been diagnosed as iron deficient, and this despite the fact that he was eating proliferous amounts of animal flesh (doctors' orders!). Similarly, there are a great deal of flesh eaters out there who still find it necessary to supplement their diet with B12. Considering that meat is purported to easily contain sufficient quantities of both iron and B12, it seems odd that their needs are not being covered by their diets. Such cases confirm my suspicion that deficiency itself is not the real issue, but rather lack of ability to absorb, through physiological contamination from poor food choices.

Biological Transmutation

In cases where it has been supposedly proven that certain nutrients are lacking, biological transmutation steps in as one theoretical way of explaining how our needs for a particular nutrient may be met despite its apparent absence in one's diet. It posits that if a physiological body is healthy and free of toxins, it has the ability to basically transmute certain elements already in abundance within the body into the ones it requires.

For many, this will indeed sound like pseudo-science, and indeed it may well be. I'm not claiming it as fact, although there is ample evidence out there that seems to indicate it may be. If you've never heard of it before, I suggest you Google it, and you will see there are many people who take it quite seriously.

Like I've said, I'm no scientist and as stated I neither personally need, nor necessarily trust the validity of supposed scientific explanations currently available. For my own needs, I simply don't really care for any of it.

Vitamin Tablets

By this point, it should stand to reason that I see no value in vitamin tablets. If I look at them, all I see is dead inorganic matter that I seriously doubt can do anything to genuinely help anyone. Whenever a new element is recognized and isolated, you will find some company somewhere that will take advantage of this fact and begin selling said element through clever marketing. "Yes, science has proven that you need phytosyllium for shiny thick hair, ma'am." "I guess I'd better get some of that, then." Such an ad is guaranteed to work especially well with those who, due to warped food choices, have dry, thin, dull, and brittle hair!

Sugar Issues

I find it somewhat distressing that so many people refer to fruit as "sugar." The 801010ers are particularly fond of this, proving their ignorance to the role of fruit, yet still calling themselves fruitarians. Sugar has a bad reputation and rightly so; it's a highly refined fully denatured product obtained through processing the hell out of sugar cane or sugar beets or similar. Then there's fully natural, designed by nature for us to enjoy and thrive on, fruit.

Whatever is inside fruit, bears no kinship with sugar.[87] Sure, from a denatured scientific viewpoint, it may contain fruit sugar, "fructose" as one of its constituent elements, but while that fructose is still locked intact and an intricate natural constituent within the fruit, it has none of the detrimental effects associated with sugar. Use a chemical lab to separate the fructose though, and its value will decrease to the magnitude of the synthesized vitamin B tablets. So if you are one of the many that confusingly refer to fruit as "high in sugar" then you need to seriously consider changing your vocabulary. Personally I think I'd prefer to call fruit sugars, fruit sweetness, and avoid the sugar thing altogether.

Diabetes

So even if all the doctors in the world (which thankfully is not the case!) were to advise a diabetic against eating too much fruit, then I would adamantly and rebelliously keep my stance and disagree with them all. If anything, it is precisely due to the lack of fruit (and consequent increase of overly cooked foods, especially meat and wheat) in their diet that they may have diabetes in the first place!

Hey, I warned you; I don't do science! Don't take my word for anything. Naturally distrust me. That's what I want to instill in people—a natural distrust in authority and the written word. Just because a doctor (or anyone) says something's true, doesn't make it so. Just because something is written down, that doesn't make it true. And like I said, I've no formal education. I just see things from a different vantage point and I can't make you see what I see. I cannot give you my clarity. All I can do is describe to you what I see.

Suffering from any flavor of disorder does not pre-empt/exclude one from embracing the Eden Fruitarian lifestyle. Nothing does. Not one's age, certainly not one's gender nor race, not even one's state of health. It's a philosophy meant for the entire world. No exceptions.

Teeth

I've heard it said by many that adopting a fruit-based diet would cause unavoidable damage to one's teeth. Indeed I myself have lost several teeth over the last couple of decades and I can understand how some people fear that a fruit or fruit-centered diet may be responsible for that; and internal nagging doubts remind them of all the scaremongering about too much fruit sugar and the effect it may have on one's teeth, bones etc.

The truth behind my own tooth saga is documented here.[88] Cutting a long story short, it was not due to eating solely fruit. My own teeth problems began way back as a shod omnivore. In my early twenties it became an almost normal occurrence that at each of my six monthly dentist visits, I pretty much inevitably ended up getting at least one filling. At one particular check-up, when I saw my dentist I was told that I was in need of ten fillings—nine normal ones and one root canal. I had my teeth built up on several occasions, partially for cosmetic reasons.

Fast forward to my path to fruitarianism, and I had long abandoned all desire to regularly have my teeth checked. Over time, the work done on my teeth began reaching its sell-by date, which, mixed with my own struggles with cooked food addiction, and through eating too much unripe fruit (which I cannot emphasis enough, being not yet ready to eat will have its consequences, and likely none too pleasant ones!), my teeth slowly fell apart. For sure, had I been stronger willed, more determined, I could have avoided that issue altogether. But I wasn't and I learned my lesson the hard way.

It would have been great if I'd succeeded in reaching fruitarianism without any dental damage; I would have been able to proudly show off my full set and say, "Look! A fruit diet is fine.

"But for me, part of my journey took me through that hardship and, despite a year or more of on-and-off excruciating agony (probably some of the worst pains I've ever experienced, to the point of delirium), I managed to avoid ever consulting a dentist.

This I did despite an enormous amount of peer pressure that bad gum infections could potentially lead to brain damage and death, but I knew that I was on the right path and that eventually I would pull through. Which, naturally, I did.

Yes, it's true, I have the same aversion toward dentists that I do towards doctors; they are both part of an industry that has no real benefit from us being healthy and will offer no lasting assistance toward that goal.[89]

Like the GPs who, on the pretext of supposedly helping us, deal out their symptom-suppressing drugs (for which, similar to the street drug dealers, they are greatly financially rewarded), the dentists, short of advising us to not eat too many sweets and to brush our teeth regularly, offer us no real knowledge as to why our teeth might not be perfect. This is most likely because the industry as a whole is ignorant of the role that diet plays in our health, including the effect the standard overcooked "flesh and two veggies" diet will have on our teeth.

The medical industry is one of the world's largest money-spinners and would have to seriously restructure and minimize itself, should we all turn around and eat what we were supposed to eat.

What makes me so sure that it wasn't just eating predominantly fruit that caused my teeth to fall out? Well, that's easy.

1. Since getting through that period of time with very weak teeth, my teeth have, if anything at all, strengthened themselves within my mouth (clearly, had the issue been fruit alone, my teeth would have progressively deteriorated and would still be an issue and causing me problems, which they are not, despite all those nasty, beastly fruit sugars and the increasingly less frequent brushing).
2. My teeth were already in a very weak state prior to my unripe mango adventure, their health having been compromised from my previous cooked food diet, and the ever-present assistance from what I once believed to be well meaning dentists who had already given me a mouthful of countless fillings.

Of course, the fillings are still an issue, as, since adopting my philosophy of non-interference and letting my body heal itself, some are still intact. And occasionally, one might reach its use-by date and disintegrate or fall out, but I don't mind.

I am glad to see the back of them! I've been told I should have them removed, as they mostly all contain mercury and there is the issue of poisoning, but I am sticking to my non-interference method and intend for things to go their natural course. Besides, I feel such an invasive procedure as mercury filling removal may do more harm than good.

Also, I believe that the body, if left to use its own inner wisdom, may possibly grow a third set of teeth; there are documented cases, although rare. If this should happen, all well and good; if not, no worries. I'm not holding my breath.

Apart from the cosmetic issue of missing a few teeth, I really don't make a big deal of it, and the remaining teeth I have do their job in regard to chewing my food.

Different Body Types

Back in the 1990s, while living at El-Faitg in the French vegetarian/vegan community of Douceur et Harmonie, up in the mountains close to the Spanish border, there was this elderly woman there, the founder of the community, and her name was Viviane. If she's still in the same body today, then she would surely by now be pushing 100. Viviane had this theory that we all have different physiological/nutritional needs and that some of us are just unable to live well on an all-raw vegan diet, let alone on solely fruit.

I recall she wrote about it in a small quarterly newsletter she edited and published at the time, making a comparison to how the horse chestnut tree and the sweet chestnut tree, although looking similar, had very different needs as far as their soil was concerned (one preferring it more acidic, or something similar). She said the same was true for humans in that we may all look similar, but our nutritional needs are actually different.

At first glance one can be forgiven for following her line of thinking, but her logic was clearly fundamentally flawed, and as much as she may have been right that horse chestnuts and sweet chestnuts look similar, they are nevertheless two completely separate species. It's like she was comparing pandas to polar bears. Had she compared a sweet chestnut tree to a sweet chestnut tree, then she would have seen that they all had the same basic requirements.

Similarly (sort of), ancient ayurvedic philosophy divided humans up into three main body types (or doshas, as I believe they called them),[90] each with different nutritional requirements (pitta, vatta, and kapha). The more modern version, especially since the advent of the book Eat Right For Your Type,[91] divides people up depending on their blood type. The theory goes that people with "Type A" blood are best off as vegetarians, while people with "Type O" blood are natural born zombie flesh eaters. "Type B's" are somewhere in the middle, needing cheesy yogurt lasagnas no doubt.

But I propose that they are all wrong. We are far more similar than any of these philosophies care to admit. Other mammal species also have different blood types, and yet no one would try and tell you that certain dogs should eat certain foods and others a quite different diet, or that some cows should eat grass but that others would be better off eating predominantly alfalfa! Of course at any given point in time, our individual requirements may differ, but overall we are all fundamentally physiologically alike. Our stomachs and hearts are generally in the same places, as our kidneys, liver, and lungs are all serving the same anatomical purposes.

What the proponents of "diet types" are seeing is apparently different types of people. But these apparent different types are just different bodies having been consistently abused in different ways.

Once the body returns to a truly good state of health, it will become clear how those seeming differences were nothing more than self-inflicted, and that we are truly all fruitarian. Nothing will sustain us, maintain us, nurture us, and help us evolve as a species more than returning to a fruit-centered diet.

The Ethical Food Tower

We've all seen the "Food Pyramid" chart, on a classroom wall or in a doctor's waiting room. It's one I believe is fundamentally flawed, as it promotes foods none of us should be eating. I propose therefore a new concept. An ethical food tower.

Unlike the traditional food pyramid we are likely all familiar with,[92] the tower differs in that only purely raw ingredients are listed. Thus you'll note the obvious absence of foods such as breads, cakes, pizzas, chocolate, and ice cream. Other than that, although visually there are some similarities between this tower and the many varieties of the traditional food pyramids in existence, the predominant difference lies not in the contents but in how the tower should be read and interpreted.

Designed by the US Department of Agriculture and dating back to the 1960s and '70s, the basic principle of the food pyramid is that we eat predominantly from the lower rank and proportionally balanced from the others. The overriding focus of this new proposed chart is one of ethics. Instead of offering a guideline or rough percentage of what foods one should eat, it shows the order of crassness of the foods, with the lower ranks being more neanderthalic in nature, which are embraced more by the dark side, and the summit being fruit, the highest of all foods and the food we should all aspire to be eating. The goal is to eat as high in the tower as possible! The tower supposes that all this food be eaten in a 100% raw (ripe when appropriate) natural state. Once the cooking process is included, things get much more complicated, and additional layering would be called for, with a great deal more potential for overlap. The tower looks roughly like this (see page 126):

Ethical Food Tower

You will note, that the bottom layer contains the edible(?) flesh portions of animals—every mammal, fowl, and fish you can name, etc. Slightly above it, with some overlap not made visible (I'm no artist), I stick the so-called byproducts of the animal flesh trade and other items that originate from deliberate animal abuse, i.e.

Fruit!
Greens
Nuts & Seeds
Tubers
Root Vegetables
Grains
Animal Byproducts
Animals

dairy and eggs. The overlap is diffused and not altogether clear. The next layer up are the grains, which, due to their environmentally destructive mono crop origins and need for heavy post-harvest processing, should never be considered ideal foods. Move up a layer again and you'll see the root vegetables, carrots, parsnips, turnips, etc.—things that necessitate the direct, unavoidable, and intentional killing of plants.

Above that, but again with a slight overlap (depending on production methods), the tubers like potatoes, yams, sweet potatoes, and Jerusalem artichokes—things that on a small scale can potentially be harvested gently from the ground with minimal harm to the plant, and if care is taken the food can be eaten and eyes cut out from the potato or similar to regrow new plants.

Moving up the triangle one more notch, once more with ever the potential for overlap, I would place nuts and seeds, but think they should only ever be eaten if truly fresh, thus with the instinctive stop still intact. Anyhow, above the nuts, I'd place greens; again, there is definitely room for overlap, as method of growing and harvesting surely plays a role, as with small-scale home gardening, many greens can be harvested in a far less destructive manner than traditional farming methods allow.

Then, right at the top, is the fruit.

Truth be known, the tower is far from finalized. Clearly I've missed out some of what gets eaten, like crustaceans, snails, insects, grubs, and the like, which I'd probably place in that vague area between layers one and two and three. Then there's seaweed, algae, and mushrooms, which I suppose should be toward the top end of the tower somewhere. Further precision could be made within each layer too (for example, tree fruits being higher than the likes of tomatoes and cucumbers), and then there are other factors like the proximity of the food, and transport issues that may mean rearranging things a little. For simplification, let's say the interpretation of the tower is suited purely for foods that are all 100% organic and locally grown.

Clearly, if there is no fruit around and one must eat other things, then move down a floor or more, as far as you feel comfortable, with the understanding that the bottom layers are really no-go zones. Of course, the ultimate goal is for us to settle in environments where the foods we are comfortable with are readily and affordably available, or to bring that environment unto us.

It should be abundantly clear from the layers that I am not equating the uprooting of a carrot to the murder of a lamb, and if you wish to you can add a bottom layer to the pyramid with humans too—the cutlets, shanks, chops, and flesh for the cannibals amongst us. In doing so, it can then also be abundantly clear that I am not fully equating the murder of an animal with the murder of a human. This should appease all those who might think I am putting the welfare of animals above that of humans. I think undermining either is an abomination!

If you can cut out the bottom layers, including the grain, then you are already making fine progress. Let your spirit rise upward with your meals.

Mono Eating

I should make it clear to begin with here that although many of our meals are basically mono, we do also regularly mix some fruit together, and although I can see there are clear benefits to eating mono, in the grand scheme of Eden Fruitarianism I don't tend to place a great deal of emphasis on doing so. If you're eating up there on the top of the food tower, and have an understanding of the basic concepts outlined within this book (there's more controversy to come, so hold onto your hat and finish reading before you nod your head in agreement!), then congrats to that!

The reason for this lack of emphasis is that Eden Fruitarianism, as I've already clearly stated, is NOT a diet but an ethical choice, and once you see that and have moved up onto the top floor of the tower, it is a matter of personal choice as to whether or not you refine your diet yet further (a personal growth choice that Eden Fruitarianism heartily encourages!).

If mono calls out to you, then I can see there are advantages of eating that way at times. It should be especially made clear that fruit is superior above other food. Try flaying the skin from a cow and biting into it. Now try peeling a mango and sampling that. Which do you prefer? Try getting on your back under a lactating goat and drinking directly from its udder, and then try picking some fresh strawberries and eating them. Then compare the two experiences. Chew your way through a plate of raw wheat, and then one of fresh grapes. Face a bowl of diced carrots, and then an equal amount of cherries. Which of these choices offers the most pleasurable experience? A plate of dandelion leaves, or a plate of durian?

My belief is that by facing our foods in this unbiased way, it will eventually always be clear that fruit is more colorful, more flavorful, and overall more enjoyable an experience.

Because shod omnivorism has lost sight of this mono simplicity (which clearly has its merits), you'll often find people eating a complex amount of foods simultaneously, confusing and overtaxing the digestive system, thus slowing it down unduly and rendering the body unnecessarily sluggish.

If we (Květa and I) mix fruit together, we tend not to get any more complex than tomatoes, cucumbers, and avocado as our often-regular afternoon meal. Sometimes there might be zucchini in there too, or capsicum, but rarely anything more complex than that.

Fruit's Not What It Once Was

People find many objections to eating fruit. One common one I hear is how fruit has been changed so much over time that it is no longer as it should be, that it is often mineral deficient or has absorbed excessive amounts of chemical pollutants through the air.

I don't think anyone can deny that all the above may at times be true; however, the one thing one should see is that of all the foods available, none of them are quite as they should be. These issues and similar ones can be applied to any of Earth's food. If you think that fruit is sullied and that pig flesh isn't, then you are deceiving yourself.

Despite imperfections, fruit is still clearly the best choice for all involved. It is our duty to recognize that and to do our utmost to help restore the planet to its potential splendor, gradually healing the soil to once more be healthy and rich, and for its fruit to be purified.

I often think this whole "demineralized fruit" issue is mostly scaremongering. If it is true, I have not noticed the effects of that demineralization, and although any of the undesired chemicals that happen to be present are surely not ideal, fruit contains enough cleansing properties of its own to help ensure they don't linger too long within the body but are pushed through with the normal physiological cleansing process.

Water

There exists a popular thought. It's been around for several generations and, like many thoughts that have been around a while, it is often quoted as fact. It is claimed by many that the human body needs to drink eight to ten glasses of water a day in order to maintain one's health. Where this idea originated is not something any nutritionist can answer, even the ones who will tell you it's correct!

Of course, I don't think anyone will claim it's a hard and fast rule—more like a basic rule of thumb guideline, based upon a variety of factors such as size, gender, age, physical condition, climate, activity level, and, most importantly, diet!

So although there may be some truth to it, in that an average person of average gender, age, and health on an average diet may require something approximating eight glasses of water daily, the same can certainly not be true for someone on a truly raw or Eden Fruitarian diet!

By its very nature, the Eden Fruitarian diet is high in fluid, with many of the fruits eaten containing 70%, 80%, 90% or more liquid, so reducing one's need for water to a bare minimum is doable. Indeed, most days I have no need at all for water. I would be over exaggerating if I said I drank an average of six liters a year. Květa does not drink any.

Although a basic guideline for water intake may likely serve its purpose for those on more conventional diets, where the high water content of most real foods eaten is generally systematically removed first through the heat process, the idea of taking a certain arbitrary quantity of water on an otherwise 100% raw vegan diet (i.e., no nuts or dried fruit either!) seems rather foolish to me, rather like the idea of taking in a certain amount of calories daily. I think once one understands what the real food intended for us is, and makes positive steps to return to such a regime, then we will all begin once more to trust the body's inbuilt wisdom. Listen to it and it will tell you if you are genuinely thirsty or hungry or not. You don't need to pay a so-called diet guru to tell you what and when to eat and drink. You just need to understand what constitutes a real food!

Children

One should not think it is necessary to wait until adulthood before embarking on the Eden Fruitarian path. Its benefits can be reaped regardless of age. There is no better choice we can make for our children than to begin them along its path. The longer we wait to do so, the more arduous it will be to reach that path. Don't take my word for it; if you are not convinced, go ask your GP.

Seriously though, my belief is that Eden Fruitarianism is a path suitable for everyone—not only suitable, but ultimately yearned for by the universe and thus, as such, we are never too young nor too old, to begin upon it. From the vantage point of the path, all it takes is an observation of small shod omnivore children to understand this. How quickly they are able to get sick and detox. And how quickly they can recover! This will likely seem paradoxical, but their quickness towards sickness is a sure sign that their little bodies are being healthy and vibrant. We may ignorantly lament over their suffering and blame other parents and children for passing on their bugs and sniffles, all the time being oblivious to the underlying fact that it is really we who are jeopardizing their health by encouraging them to eat unsound food choices.

Persevere on the shod omnivore path and at a certain point our children will appear healthier and no longer be constantly struggling, or fending off childhood bugs. This has nothing to do with them being healthier though but more to do with the body gradually resigning to its often-lost cause of struggling against foods it was never designed to process. Of course, with our bulging waste lines and warped senses of what constitutes good food, we are rarely able to recognize any of this.

If we are fortunate enough to understand this, we are then faced with the issue of guiding our children back on the right path. This will not be straightforward. One can rightly expect protest, complaint, and outright tantrums. One has to understand they are coming off a drug and therefore they will experience symptoms of "withdrawal." Be there for them. With sound help, guidance, and above all patience and understanding, they can get through it.

Even if you, as a parent, still struggle with the addiction of cooked and overly processed non-foods, once you have understood the detrimental nature of such foods, it will be your moral obligation to not let your children follow the same health-destructive path. Think of it as if you were a smoker, struggling to quit. If you yourself can't find the strength and commitment to break free of the habit, would you think it acceptable to let your child smoke too? I am not suggesting you forget your own struggle, because unless you too can follow your convictions, there is little chance that your child's success will be lasting.

Environmental Issues

It is a sad fact that this planet we call home has been and continues to be violently trashed. The predominance of that trashing has been through the hands, thoughts, and deeds of our very own species. Guilty as charged. Our consumer habits necessitate exploitation, whether it is slave labor; the innocent imprisonment (and ultimate bloody slaughter) of other sentient species; the widespread destruction of arable land and forest; the pollution of air, waterways, and, oceans; or ultimately our own individual health. Of course, none of this is necessarily immediately apparent, but pretty much everything purchased in any mainstream outlet store has its impact.

For many years deforestation has been ongoing at an evermore-alarming rate. By some reports, as much as or more than

70% of the world's forest areas have been flattened within the past 200 years, much of that within the past few decades. It's like the planet is now forced to breathe with less than one decrepit lung!

At first glance, one might think, with the overwhelming negativity of it all, that there is no escaping any of it. Indeed, short of totally going bush and growing one's own fruit through completely sustainable methods, it is currently virtually impossible to avoid some contribution of negative impact.

An individual stepping onto the path of Eden Fruitarianism cannot expect to see immediate global changes. That would be just unrealistic, and as much as many might judge the idea of Eden Fruitarianism itself as unrealistic, I believe there is a profound realism that supports it. One should not be overly concerned with what everyone else is doing but focus purely on lessening one's own carbon footprint.

As a consequence of fully embracing Eden Fruitarianism, suddenly many of one's once-popular consumer items will no longer be of appeal and thus, by the very nature of things, our impact will lessen. I am not talking purely of food here but of pretty much everything else on the supermarket shelves that isn't in the fresh produce section. No more cans, no more excessively packaged cereals, no more cleansing products, no more oils, no more conditioners, no more denatured junk foods. The list is longer than most pieces of string.

Indeed, it is safe to say that most commercial businesses will ultimately reach their demise once Eden Fruitarianism has been fully understood and adopted by the majority.

This is a good thing!

Agriculture

I wish to make clear once more that, similar to veganism, Eden Fruitarianism is not a diet. This is what distinguishes it from other flavors of fruitarianism. Unlike the traditional omnivorous or vegan diets that are based primarily on annual crops—predominantly cereals—a fruit-based diet is predominantly based on perennial agriculture, primarily fruit tree growing. As Eden Fruitarian consumers, we will be demanding more fruit, and more trees will need to be grown.

Being reliant on annual crops means heavy reliance on fossil fuels for land cultivation. It involves widespread loss of wildlife habitats and annual unintentional slaughter of innocent animals that get caught up in the heavy, brutal machinery used to turn the Earth. The Earth suffers from this practice, which is, to say the least, highly unnatural. For ethical vegans this should be reason enough to at least forsake the third floor of the Ethical Food Tower!

In nature one cannot find a naturally grown field full of one particular plant species. Mono crops are devastating the planet. Growing large mono crops to feed the masses involves a constant struggle against diversity of plant life.

The alternative of not relying heavily on pesticides, herbicides, and other chemicals is even more labor intensive, resulting also in higher produce prices.

Fruit tree growing does not involve the Earth being disrupted each year by heavy rotating ploughs and the like. That is not to say fossil fuels and chemicals are not used in fruit production, far from it. The amount and impact is, however, vastly reduced. And I mean vastly. Fruit trees are currently also grown in large mono orchards—field after field of mangos, etc.

Ultimately, I clearly believe this can also be improved, though the environmental consequences are far less than the desertification tactics of annual mono crops.

It takes up to 15 thousand liters of water to produce 1 kilogram of beef compared to only 2.5 thousand liters to produce 1 kilogram of rice, and significantly less still for tree-based agriculture producing 1 kilogram of fruit.[93]

Land Animal Farming

The UN has released a report titled, "Livestock's Long Shadow."[94] They explain how the animal exploitation industry—major suppliers of unhealthy addictive foods—is one of the most significant contributors to the most serious environmental problems the Earth is currently faced with.

The solution? Move away from animal farming and mono cereal crops, and begin focusing more on horticulture and small diverse fruit tree orchards. Not only will this help restore the planet to its splendor, but also it will ultimately save and prolong lives and serve to awaken awareness.

In addition, animal industries habitually negatively impact biodiversity by causing loss of native habitats, and introducing non-native species, which leads once again to unnecessary competition for both food and water.[95]

The solution is clear again: to use land to plant more trees, creating more biodiversity and offering a long-term solution to food production through environmentally sustainable fruit-growing techniques.

Additional detrimental impacts animal raising has on fresh water supplies (all of which can ultimately be avoided by ditching dark-aged flesh-eating habits) are that livestock often trample and destroy river edges and pollute waterways. Native vegetation is systematically destroyed to make way for mono crops or pasture land that will ultimately cause soil erosion and affect rainfall.[96]

The manufacturing of leather and other so-called animal byproducts are major sources of pollution, with the reliance on formaldehyde and other toxic chemical pollutants.[97]

In the US, animal factory farming pollutes more waterways than all other industries combined.[98]

A large chunk of greenhouse gas emissions can be attributed to conventional animal farming techniques. Take into account the fuel and energy consumption such farming methods rely on; the statistics are surely even more disturbing. Animals raised for food in Australia alone produce about 3.1 megatons of methane annually.[99] (Methane has a far greater global warming potential than carbon dioxide.)

Believe it or not, grazing takes up nearly 50% of the Australian continent's arable land. Roughly 380 million hectares![100]

Instead, we could be planting more intermixed varieties of fruit trees. Eating more fruit and ultimately cutting the meat, wheat, and cheesy dairy from our diets is a very effective way for individuals to make a very real difference in reducing global warming.

Additionally, livestock animal grazing has a direct impact on the environment by compacting and acidifying soils by increasing to unsustainable levels the volume of manure, and other byproducts, effecting both land and waterways.

According to the CSIRO[101] and the University of Sydney, 92% of all land degradation in Australia is caused by animal industries.

Increased numbers of agricultural animals, over-farming, and overgrazing has led to vicious cycles of deforestation, erosion, and habitat destruction. Eventually this will lead to starvation prompted by the disappearance of plant food sources.

Fish Farming

Meanwhile, out in the ocean, innocent fish are being caught by the tens of thousands, all to satisfy human gluttony for their zombie flesh-eating habits. The consequences are that these modern fishing methods have depleted populations to such a level that the industry is now targeting deep sea fish, species not previously taken.[102]

Not only are fish populations being forced into a spiraling decline, but also destructive fishing techniques including dragnets, long lines, purse seine nets, and drift nets are destroying large parts of the ocean environment in the process. By unintentionally capturing other sea animals such as non-target fish species, whales, dolphins, turtles, seals, and sea birds like the albatross, many of these species are facing extinction due to fishing. Insatiable gluttony for food with eyes has created vast areas of the ocean void of desirable fish.

All this mindless destruction while the only real food that encourages a healthy body, mind, and planet is fruit, which goes virtually unrecognized for the solution it harbors.

Fish farming is equally insane, creating concentrated fecal contamination in specific areas of the ocean and rivers, and promoting the rapid spread of disease and parasites to both captive and wild fish populations. Even more bizarre is that farmed fish are even fed fish! Five kilograms of wild fish is needed as feed to produce one kilogram of farmed fish and up to twenty-two kilograms of wild-caught fish needed to produce just one kilogram of farmed tuna.[103]

Crikey, that was one boring screed! I just wrote it down and read it back to myself and all I could hear was, "Blah blah blah." In the future do your own Google research, won't you?

I could continue listing insanities, many of which, probably together with yet more mind-numbing ones, are available on the

Internet, but what would be the point? The only real solution is very clear to me. Leave behind destructive habits reliant on the slaughter of fellow beings and their homes and instead plant fruit trees and eat their luscious offerings—the most superior of all nutrition and the only truly environmentally friendly and truly karmically free food. FRUIT!

How, When, and What to Eat

The rules are so, so simple really. Once you've pushed through the fear and uncertainty, the only thing you really have to understand is the valuable role and potential of fruit to restore peace to the world. Once you have that, all you then have to do is present yourself with only fruit to eat and trust your own body to intrinsically know what it needs. Sit down, look at your food, smell it if you want, feel it, grok it. Hungry for a cucumber? Then eat one. Feel like tomatoes? Eat as many as you want, as often as you want, whenever you want. Still hungry but want something different, and avocado looks appealing? Then eat one or as many as you want. Feel like mixing them all together? Go for it!

Don't make the 801010 mistake of ruling avocados as fats. Avocados are a fruit. Lumping avocados in with nuts is a flaw in their pseudo science. As stated, all commercially available nuts are cooked, thus their overall effect on the human physiology is bound to be comparable to that of cooked food, and thus ultimately be detrimental. Avocados are neither nuts, nor cooked, nor heat-treated, so their effect is very different and not comparable to that of nuts. Don't be afraid of them; be afraid of the pseudo-scientists who would rather blind you with their half-arsed science and bring you to a sense of fear that they alone have the complex knowledge you need to stay healthy.

Don't get caught up in calories, omegas, antioxidants, free radicals, and electrolytes (things you will likely never really understand anyway!), or force yourself to eat more than you need. Become yourself a free radical. Don't get caught up in drinking copious amounts of water even though your body doesn't feel like it, just because some snake-oil salesman tells you that you should. Stop looking for other people to tell you when and what to eat, and just go with the flow and conviction that fruit is your food. Analyzing which microelements are in each food doesn't matter; just trust your own physiological wisdom to guide you! If you feel like eating banana and tomato mashed together, then go for it!

Unless you are underweight, you'll not put on more, even if you eat fruit all day. If you are fat already, then you will lose weight until you reach your ideal one, which by society standards will be considered skinny, but onlybecause society is fat. If you look at a fruitarian diet with the preconceived notion that you'll never be able to succeed with it, then you will be setting yourself up for failure. Seriously give it a shot and you may be surprised to discover just how little fruit is needed to keep one satiated and how a simple large glass of orange juice can give us enough energy to keep us going for longer than one might think.

There are simple rules: Eat when hungry, drink when thirsty, rest when tired, sleep when sleepy, shit when...shitty? (It's a known fact that being constipated can make one angry!) Don't eat and drink for tomorrow. You're not a camel living in a desert!

To avoid issues of not getting enough, avoid situations where there isn't an easily accessible plethora of fruit. It's no good planning to do a bike ride or long walk and force-eating and drinking everything you think you need before you start out.

Locally Grown Produce

It should be clear that the closer your food is grown to where you live, the better it will be for everyone and everything involved. Currently there are all manner of things fruit, and food in general, must endure when transported vast distances, the least of which being the fuel used. Should the food cross national boundaries, far worse treatment often awaits it. Such treatment is known as phytosanitary measures,[104] and it frequently involves extended processing of the food even before it leaves its farm of origin. The supposed goal is to prevent importation of pests through food trade, and this will often involve irradiation, dipping the produce in an insecticidal soap or light paraffinic oils, and further treatment with fungicide (to control post-harvest rot). As a rule, these are all processes NOT undergone when the food is grown and bought locally.

All the more reason why the committed Eden Fruitarian should make sincere long-term plans to relocate to where the fruit is grown!

Shit Happens

Once we have had our meal, the inevitable happens. Nature calls. It is generally said that human waste is not a good addition to the garden. The main reasons you'll see quoted are health issues, and there's often this feeling that it's much more dangerous to use human manure than any other animal's in your garden.

Well, I think this piece of folklore has been disproved in many places, as human sewage is sometimes treated and used as fertilizer on farms, but if there is a shred of truth to it, then the truth is centered on the stinking feces of the cooked shod omnivore. The waste of a fruitarian is comparatively much cleaner and far less intensely smelling and thus more beneficial for the Earth. Eden

Fruitarianism recognizes this and encourages us to stop flushing such waste to unknown fates with the sewage works and instead take responsibility for all our waste, recycling it as best we are able.

Consider also using water to clean your behind instead of toilet paper. After all, if you had shit on your face, then you wouldn't just wipe it off with a tissue and think, "It's clean now."

Fitness and Exercise

Any quick Google search will show that many of the batches of novo fruitarians out there have a strong fixation on some kind of fitness regime. This may be daily sessions at the gym, a strong desire to run fixed or variable distances daily, or some ingrained desire to forever push the limits of their bicycling endurance.

Such exercise is not altogether mutually exclusive with Eden Fruitarianism, but preoccupation with alpha egos, six packs, abs, biceps, triceps, steel buns, competitively outperforming one another, and an often-morbid fascination with apex predators and survival of the fittest. There is recognition that the best form of daily exercise has goals beyond pushing oneself to beat individual or group records.

Why run around the garden expending energy solely for one's individual fitness, when one could be using that energy to dig holes to plant new trees, to climb trees and harvest their fruits, to weed and mulch around those trees, or to prune the dead branches from them?

Such actions can be both rewarding and meditative, use up just as many calories (whatever they are?), and are far less likely to result in physical injury than the harsh running on concrete, the breathing of conditioned air in gyms, or the dodging of traffic on one's push bike.

Personally I don't see pushing oneself to the limit as being ultimately that beneficial, and it's more likely to cause injury than more gentle techniques that can be employed like yoga, walking, swimming, etc. (or, again, a healthy relationship with one's garden!).

Weight Loss and the Biggest Loser

Whether or not you are into daily exercise purely for the goal of personal fitness, weight loss is a virtual inevitability on the Eden diet. Weight loss has pretty much always more to do with diet than exercise. I'm not sure if you've ever seen that TV show, The Biggest Loser. Květa and I used to watch it some years ago, back when we still had a TV. The basic premise of the show is that they get a whole load of obese people, split them into two groups (a red team and a blue team), and make them compete to lose weight, with one member being voted off from each team each week. They teach everyone to eat healthier than they were, while meanwhile turning them into trained athletes with rigorous daily exercise routines.

The Green Team

We used to talk of how it would be great to have a third team, the green team. This team would be put on a cleansing fruit juice cure and an Eden Fruitarian diet. Instead of being taken through boot camp comparable training, they would be merely given gentle daily exercise routines like swimming, walking, yoga, and gardening. Not being pushed to the limit until throwing up but just doing mild movements.

While the blue and red teams are being taught about calorie counting and good and bad cholesterols, the green team, under the shade of some large fruit tree, are being given lessons in empathy and compassion, the precise lack of which being the real reason they find themselves in such awkward body predicaments!

Instead of being out in the hot sun, or in some air conditioned gym exercising relentlessly, we would take them on field trips, let them witness the birth of a calf, let them see how the mother gently cleans it afterwards, and then let them see the agony of separation. How both the calf and mother suffer and call out for each other. Let them witness the conveyor belts of chicks waiting to be sexed and show the destiny of both genders. Teach them about branding, tagging, and debeaking. Let them see the factory farms and the lifelong incarcerated conditions of animals destined to be gluttonously feasted upon. Let them witness animals as they are prodded and shocked toward the delivery trucks where they often spend hours packed in overburdening conditions on their final arduous journeys to the slaughterhouse, and let them see the slaughterhouses and hear the screams of animals defenseless to change their destinies.

Show them the same animal species in unfettered realms and teach them how each one is as individual as any cat or dog (or you or I). Let them see how every living thing desires a life of freedom and relative comfort void of fear and threat. Take them to see mono crop farming and tell them how it is a constant struggle, an inevitable losing battle against nature. Let them really see what their dietary habits entail, so that they may later make more informed choices and live with clearer consciences and minds, with more bliss and less ignorance.

We'd tell them to take their shoes off and go for a walk in the grass. Take them to beautiful nature spots and let waterfalls massage them. Teach them that being fat is never ever due to genes, nor water retention, nor being big boned; that even glandular issues can be rectified; and that everyone who's overweight can and should lose weight. No exceptions!

Perhaps we're wrong, but Květa and I are both convinced that the green team could quite easily beat the blue and the red one each week. The pounds or kilos would literally melt off of the participants!

Eden Fruitarianism Continued

This is not all there is to Eden Fruitarianism. There are more aspects that merit their own section and are thus discussed in the following chapters.

Section Two

ADDITIONAL EDEN FRUITARIAN
LIFESTYLE CHOICES

*They who dream by day
are cognizant of many things
which escape those who
dream only by night.*

-Edgar Allan Poe

Chapter 6
Barefootism

Once Eden Fruitarianism is fully understood and embraced, shedding one's shoes is a natural consequence. I think we all secretly love the feel of freed feet, the flow of fresh air upon them and the unimpeded ability to wriggle one's toes, and at some point the desire to liberate one's feet supersedes any desire to conform to the dictates of fashion.

After a while you may grow to consider shoes as perhaps the ugliest, clumsiest items of clothing humans regularly don. It has become a complete mystery to me how people are so attracted and attached to them, some to the extent of developing a complete shoe fetish, obsessed with forever widening their collection, with dedicated wardrobe space and having a pair to match every outfit. I fully sympathize with the ancient Greeks, who viewed footwear as self-indulgent, unaesthetic, and unnecessary.[105]

It should be clear that the wearing of footwear volitionally is an exclusively human characteristic. I can get the practicality of shoes worn under certain circumstances. During sports, like skiing or football, or in temperatures experienced during winter months, or hot sands in the heat of the summer, or the sharp rough rocky terrains of some areas in nature. But the more one habituates one's feet to the freedom of being barefoot, the tougher they become and the more they can handle such extreme circumstances.

It feels good to have the Earth between my toes and to no longer have my feet restricted to confined sweaty spaces. I no longer even own any footwear. I stopped habitually wearing them back in the early 1990s, and have not worn any since boarding a plane back in 2006 and then only while walking through customs and onto the plane!

I read once that original Australian Aboriginal peoples, with their ever-nomadic lifestyles, would habitually be barefoot, and were they to come across terrain too difficult to pass through comfortably, would sit down and make themselves footwear from woven reeds or grasses or tree bark, etc. They would walk through the rough terrain and then discard them, letting nature once more claim them. I think they knew quite well the importance of keeping contact with the Earth through the soles of one's feet.

Nowadays shoes are fashioned from any manner of exploitive materials and through many a manner of exploitive methods, the most expensive from the flayed hides of butchered animals and, also more commonly these days, from destructive petrochemical byproducts.

Confronted with the notion of not donning footwear, many may immediately prejudge the idea as dirty, unhygienic, and a sign of poverty. Although these are unfortunate aspects regularly connected with the practice, I want to make it clear how barefootism from a fruitarian perspective differs. It may be true that the barefootism associated with the city homeless, often those with mental health issues, or drug or alcohol problems, or just outright poverty (caused by the underlying mentioned issues), is a reluctant situation, where practitioners have regularly also lost some or all sense of personal hygiene. Fruitarian barefootism, on the other hand, is a joyous lifestyle option based on complete freedom of choice, spiritual grounding, and a strong desire for mental and physical personal cleansing.

Years ago, back when I myself was a shackled and shod omnivore, by the end of a normal day's play or work, I would remove my shoes, unleashing a terribly unpleasant foot odor in the process. There were times when I would carry one or even two extra pairs of socks with me, and change them throughout the day. My feet would get sweaty as well as stinky, and they often felt quite uncomfortable after a day of shoes. This is by no means an uncommon problem; shoes of any shape or form will all gradually and inevitably lead to foot health consequences of some form or another.

True, walking barefoot may mean the occasional glass shard or thorn splinter and, if done with the wrong frame of mind, may sometimes result in an odd cut or so, especially for newbies, but over time, as the soles of our feet regain their natural supple firmness, such occurrences will always lessen. Besides, not using footwear will also encourage us to have a greater awareness of what's in front of us, and how and where we place our feet, thus rendering it far less likely for us to step onto, or into, something unpleasant or living. Barefootism quiets one's inner thoughts, clears one's mind, and helps one to focus on the here and now, allowing one to pay attention to every step one takes.

I have been cautioned on numerous occasions about walking barefoot through the bush; aren't I afraid of snakes? Shouldn't I wear shoes just in case? Such admonishments, although naturally well meaning, also show a clear lack of understanding of barefoot benefits. There was this one time, at band camp, that I did so happen to stand on the tail of a snake hidden in the grass. Because I was barefoot, I immediately felt that my foot was pushing onto something living, and thus immediately recoiled to see the snake slither away from me. Had I been wearing the heavy walking boots people get so obsessed with, it is quite likely that I would have stepped with full force onto the animal's tail, which could then likely have hurt and enraged it enough for it to part with some of its venom. As it happens, the majority of snakebites are between the ankle and knee, so unless one wears knee or thigh-high boots, it's unlikely that in such situations snakebite could be avoided.

Letter to Librarian

Back in 2002-ish, I found myself in Glastonbury, England. I'd heard the local library was a good place to get free Internet, so I would occasionally visit for an hour or two. One day, while visiting, one of the librarians approached me and asked me politely to please not come in barefoot any more. She said it was dangerous, there could be glass, that feet were dirty, and that small children sometimes crawled on the carpet.

I felt this was an odd request, considering I was quite sure I had seen other barefooters in there previously (Glastonbury probably has the highest percentage of barefooters in England), but decided not to object there and then.

Instead, I left the library and wrote her a letter:

Dear Library Lady,

Hallo. I don't want you to feel bad about asking I wear shoes.

I accept that your request, although I believe personally that it was based on false knowledge, was well-meaning.

Truthfully, I have not worn footwear since the early half of the 1990s. I have travelled across much of the European continent by bike and on foot, as well as various other parts of the world. Rarely have I, in the course of that time, experienced anything more than the very infrequent splinter.

I don't own my own footwear and have borrowed the sandals I'm wearing, in order to walk across eight meters of carpet that we both know is well cleaned and without shards of glass. Besides, I can assure you that by being barefoot, one has far greater awareness of where one places one's feet.

With the cleanliness issue: In comparison to the soles of most people's footwear, my feet are far cleaner. Feelings of dirty, sweaty feet are based on feet cooped up inside socks and footwear all day, which, in such clammy, awkward, restricting environments, will naturally be a thriving breeding ground for bacteria and manufacture undesirable, unpleasant odors.

Walking barefoot is part of my pilgrimage that has brought me to Glastonbury temporarily. It is about grounding oneself and regaining contact with Mother Earth. It is about permanent reflexology. The health of feet will gradually deteriorate over a lifetime of footwear (ab)usage. Tinea between the toes and other bacterial fungal infections will almost inevitably develop, and most people will have deformed feet and other foot problems in their later years caused by shoes.

I am not drunk, loud, nor otherwise obnoxious. If I choose to let my hair grow unhindered and my beard go wild, this is basically Essenic in nature and I trust my appearance has not caused prejudgment of my character as is sometimes the case.

For these reasons—and also because I have recently spoken with several other people who regularly visit the library sans foot attire, none of whom have ever been asked to don shoes—I would kindly ask that you reconsider allowing me to remain barefoot.

Kind regards, in love and light,

Mango

The next day, I borrowed a pair of sandals, went to the library, put them on in the entry hallway, and went over and sat in my favorite corner. I read some, then as I left, I handed her the letter.

I didn't see her for a week or two, but I went in there one day and she came across to me and said she'd like a word.

She told me that she'd shown the manager my note and they'd both agreed that I could come in barefoot. I was so pleased. Even sitting down with sandals on for that short time, my feet were feeling increasingly uncomfortable.

Permanent Reflexology

Back when my feet would be regularly cooped up throughout the day, I can attest that, if the shoes were comfortable and cushioned enough, walking would seem to be so too; however, at the end of every day, my feet would inevitably feel wasted and worn. The reverse is true for me now. I feel every step I make—whether it be across hot tarmac, or rough bush terrain—I am aware of every step. Gravel, stony ground, twigs and unevenness, hot sands, sometimes to be sure a step might hurt slightly, certainly far more than any usual step in shoes would, but at the end of the day, my feet generally feel great. They are odorless and generally a lot cleaner than they ever were imprisoned. It is like I have tapped into free permanent reflexology/foot-zone therapy.

Chapter 7
Naturism

Nudism

Mention Naturism and likely the first thing that springs to the minds of many is bare bottoms running around. There is no denying that nudism is indeed an intricate part of what naturism is about; however, the two terms are by no means identical in meaning. Unlike nudism, which is basically the practice of going around naked as much as possible, naturism implies also a deep desire to seek out and enjoy more natural settings. To swim in clean, fresh flowing water, rivers, streams, lakes and ocean. To be massaged by waterfalls, to lie naked in the sun and feel its gentle caress. It implies a desire to escape civilization, all the fumes, pollution, and concrete; and to enjoy and live a more back-to-basics lifestyle.

Nudism does not imply being constantly naked but is instead about dropping preconceived society-influenced notions of nudity being in some way shameful. It's about letting go of the fear and stopping being so clothed-minded. If you break a bone and have it plastered to prevent further injury, you'll know the skin under that cast, through being starved of the freedom of fresh air and sunlight, will slowly atrophy. The nerve synapses will diminish to a whimper of their former glory, and the muscles will gradually weaken due to strangled restricted motion. As modern day homo sapiens, there is an overwhelming tendency to shroud our entire skin palette, except for face, neck, and hands, effectively obliterating health nourishing symbiosis with the planet. We hide in cocoons, when we could be free as butterflies!

A 2003 University of Reading study entitled, "A Naked Ape Would Have Fewer Parasites,"[106] posits that "humans evolved hairlessness to reduce parasite loads, especially ectoparasites that may carry disease." Unfortunately, the restrictive clothes we don can be breeding grounds for foul fungi and bad bacterium, causing yeast infections, urinary tract infections, and more. Cinched-up belts, ties, and elasticated clothing impede breathing. Men's snugly fitted underpants raise testicle temperature, lowering sperm count and fertility. I'm sure the list continues.

Furthermore, health benefits of social nudity include stress reduction, satiation of curiosity about the human body, reduction of porn addiction, a sense of full-body integration, and developing a wholesome attitude about the opposite sex. Research at the University of Northern Iowa discovered nudists have significantly higher body self-acceptance,[107] while a somewhat similar study[108] concluded teens at a New York nudist camp were "extraordinarily well-adjusted, happy and thoughtful." It's also excellent for children to grow up free of shame about the human body. Yet another study of 384 participants[109] concluded pro-nudity students "were significantly more accepting of other religious groups and gays and lesbians" when compared to the anti-nudity students. They were also "less prejudiced towards ethnically dissimilar others."

Yes, we surely live in a prudishly warped world, full of petty hang-ups and issues, where nudism is generally shied away from, ridiculed, and frequently considered a perversion practiced solely by sexual deviants.

The True Perverts

Eden Fruitarians are able to see past such uncivilized beliefs and recognizes the true perversion inherent within such thinking. They understand nudity as an open, integral component of Eden, ultimately unashamedly innocent, and embrace the concept with the shamelessness of a child.

Clothing is a strange concept. Especially this warped sense of "smart" that accompanies it, based on what some rich Italian fashion designer may think best: men who unquestioningly and habitually tie ropes around their necks (ties) and women in starchy dresses with fleckless, overly pronounced creases. Facial growth is never allowed to develop. I always thought, and still do, that this whole idea of "smart," is nothing short of bizarre at best.

Admittedly, though it seems now to be a lifetime ago, I did try to fit in for a while. I played around with the idea, three-piece pinstripe suits, the daily telegraph tucked neatly under one armpit, with an ever-visible, half-filled crossword and similar. I think I managed that for about a year or so, before growing tired of the charade. Although I did try, I was never really one able to fully embrace the notion, and always felt like I was an imposter, looking odd at the best of times. Rather like a punked up rooster with socks. Even as a kid I used to resent it when my parents would say on a Sunday, "Put your smart clothes on." I would always do so reluctantly muttering, "But I'll get them dirty and you'll get upset with me!" The inevitable reply was, "No we won't; don't be silly!" and the inevitable outcome was that we'd go to the park, I'd end up with grass on my knees, and: "Oh just look at your knees!" What use is life if you aren't allowed to get your knees dirty in the grass?

People can freely choose to look this so-called "smart" if they desire. The concept of smartliness holds no appeal for me. Nowadays

I'm in to minimal clothing. The ultimate ideal would be to rid myself of clothes altogether, to abandon them with complete non-attachment. Climate and society dictate otherwise though, so I practice only whenever victimization is off the table.

I recall, while travelling through Malaysia in the early 2000s, sitting on a bus in the middle of nowhere in particular. It was hot out there, but judging by the apparel of the locals, one may have easily been deceived, covered up neatly with thick denim jeans and baseball caps, long-sleeved shirts sticking sweatily to their backs! How long has it been this way? Why are we all such prudes? I don't care how my body looks; I'll gladly bare it anywhere if it means eating some sun and doesn't result in ridicule, harassment, or punishment.

Rather similar to the footwear issue mentioned in the previous chapter, many so-called "smart" people mistakenly equate "non-smartness" with uncleanliness/dirtiness. This need not be the case at all, as the Eden Fruitarian is in many ways quite concerned with hygiene and cleanliness. Just not via means of soaps, shampoos, or toothpastes. I believe profoundly that cleanliness is next to Godliness, and that smartliness and cleanliness have nothing to do with each other. If only smart people were really smart and could see their insides! They are generally like heavily decorated houses that look overly sparkling and glamorous on the outsides, while their insides are rodent-infested mounds of discarded dirty old socks, covered in kitchen grease, dust balls, and piles of forsaken washing up. Ha! Hippy bathday to you!

Soaps and Shampoos

Yes, in desiring to get back to nature, it should become clear to those who give it enough thought that items such as soaps, detergents, cleansing powders, makeup, and more hold no real value. It's funny how we associate soap with cleanliness, seeing as soap in and of itself is so "not" clean. It is pretty much always highly processed and generally full of non-existent naturally occurring chemicals, concentrated in profusion. Most soaps include also some ingredients from the animal industry too (e.g., gelatin). Water sullied with soaps and so-called cleaning products dirty and pollute the Earth. Natural Earth on the other hand is fun and, contrary to what we've been brainwashed into thinking, cleansing. A good mud bath can cool you down in the heat of the day and really cleanse those pores better than almost every soap.

I stopped using soap in the mid-1980s and began instead to rely purely on water. Shortly after that, I quit using shampoo too and ceased also using detergents and other household cleaning products, including washing powders. Ceasing the use of shampoos was the most difficult. I guess the hair becomes addicted; it seems to begin craving it. But after just a few weeks, the head somehow adjusts to the lack of the presence of shampoo, and things normalize again. I really believe soap is detrimental to the skin, ageing it prematurely by robbing it of natural oils and subjecting it to whatever chemicals are contained within the soap.

I know many readers might argue this case, stating there are environmentally friendly soaps, or sometimes you might believe the only way to get rid of something is with a soap or a heavy detergent of some kind. But the reality is probably that so-called environmentally friendly soaps are just not as polluting as the other more commercially available ones. Sure, the original raw ingredients may theoretically be relatively harmless, but is the harvesting of them? Or the process they undergo? And what effect will polluted water have when returned to the soil or into waterways?

As for those times when you're having difficulties with stains, I say, think twice before bleaching. Leave the stain be. Don't let yourself be ruled by fashion or by false senses of smartliness. Every stain gives the clothing history!

Toothpastes are often employed to disguise the bad breath originating from bad diets. Eat properly and any need to cover up one's internal body odors with minty fresh pastes will vanish.

Sun Creams

We have been told how sun creams prevent skin cancer. Indeed, society encourages its use whenever we spend extended periods of time in the sun. Some refuse to leave the house without it.

My personal viewpoint is one of extreme skepticism toward this supposed beneficial usage of sunscreen. Actually, perversely enough, I really believe the reverse may be true, and that sun creams may themselves be partially responsible for evoking skin cancer. Put aside what you have been told, what is generally accepted as being fact, and ponder over the following: Sun creams contain a whole host of chemicals that are far from natural. In fact, some of the ingredients in sun creams have been demonstrated to be carcinogenic (actually promoting the devil-upment of cancer cells!).

Actually, just reading the labels makes me feel queasy, with ingredients such as isotridecyl salicylate, Octyl salicylate, and Butyl methoxydibenzoylmethane. Need I go on?

Sun on the other hand, is totally natural, and without it all creatures on this planet, likely even all life, would end. It gives us a steady supply of Vitamin D (whatever that is!), which is believed by many to be crucial for the absorption of Calcium in the body, but from a naturist, barefoot fruitarian's perspective, the most important thing we need to understand is how (providing we don't overdo it) the sun just feels plain great!

Skin, along with one's kidneys, intestines, liver, and lungs (S.K.I.L.L), is a major organ of elimination, meaning some of what we detox emerges through the skin. It is my belief that the sun's caress is of vital importance to help cleanse the skin of surface toxins. By applying oil and chemical-based sun creams to our skin, we are effectively hindering the skin from detoxing. In addition, we are actually absorbing new chemicals through the skin pores from the lotions themselves.

At some stage on the Eden Fruitarian path, you will begin to know, through increased olfactory awareness, just how toxic sun creams really are. The toxic stench of the beach-going masses who spread it sickly thickly onto their poor toxin-overloaded bodies is at times quite overpowering.

During the heat of the summer, after gradually letting more and more sunshine onto my body, I find I can be outside most of the day and rarely get burned. Sometimes I find my nose is a bit of a weak point, and will keep it covered to prevent too much from skin flaking off. Should it happen that we get burned, we just apply a little aloe vera fresh from its leaf.

We also enjoy massaging fruit peels onto the skin (especially papaya or mango!) and find that really helps the skin to not get burned too.

If you're not convinced that sun creams might cause cancer, I suggest you research further.[110]

Hair

In seeing the folly of the concrete jungle, with its accompanying peer-pressured insanities, it should make us step up and question pretty much everything. Society tries to hammer the wildness out of us and turn us all into sheeple, like drones who follow unquestioningly. One expression of that conformity is through taming our hair. We might think we are expressing our individuality through shaving it all off, ironing it flat, perming it all frizzy like, or dying it pink with blue stripes, but these are all encouraged practices that generally ultimately serve to grease the wheels of the concrete economy.

The Eden Fruitarian dispenses with such ingrained conditioning and the thought that hair styling is in some way important, and finally lets hair grow naturally, however they so choose. Along with chaetophobia,[111] pogonophobia[112] is completely absent, and if one's beard decides to reach for the bellybutton, then so be it! Make it so!

I was reading an article[113] online recently that indicated how long unhindered hair might in some way be connected to a sixth sense. The article described how during the Vietnam War, experts were sent out to find Native Americans who were blessed with uncanny, seemingly supernatural abilities to track.

A couple of American Indians were selected. What followed will be surprising to many. Once they were recruited, their natural skills and ability to access their sixth sense just disappeared. On conducting a test to find out what went wrong, the older Native American recruits said that after undergoing military haircuts, they could no longer sense the enemy; they lost their intuition powers and were no longer able to access subtle extrasensory information.

So more American Indians were recruited, and this time they were allowed to keep their hair long. When comparing the performances of men with standard military haircuts with men with long hair, they discovered that longhaired men were far more skillful. The results stated that the Native trackers should be allowed to keep their hair long.

I can't help but wonder how they may also have been affected through relinquishing contact with the Earth and being forced to wear boots!

Long hair has long been a common element of many spiritual prophets such as Jesus, Moses, Buddha, and Shiva. Likely many of us can recall, too, the story of Samson and Delilah from the Bible, when Samson lost his strength with the loss of his hair.

It appears that hair may in some way be an extension of the nervous system. If this is true then that when hair is cut, receiving and sending transmissions to and from the environment will surely be hampered.

Temple Graffiti

Our bodies should be seen as temporary sacred temples of the soul and our duty is to keep our individual one clean, both internally and externally. Internally this is done by eating the right food, which is fruit. Thus avoiding any foods that sludge and sully the interior (especially animal products and cooked foods of any kind!). Externally, bathing in water regularly will normally suffice, but all fruits can also be massaged into the skin with benefits. The skins of mangos and papaya feel especially pleasant, as does a head/hair bath in lemon juice.[114] Let them soak in for at least ten minutes before gently washing off with natural water (preferably non-chlorinated, and non-fluoridated); you will feel the benefit!

Eden Fruitarians recognize the defacing of one's external temple walls through carving tattoo patterns and shapes onto them as being somewhat sacrilegious. They have no desire to vandalize the temple through puncturing holes around, through, and into it, or to adorn it with metal rings, chains, and studs. Even the less harsh graffiti of makeup is recognized as being detrimental to the skin and thus should be avoided at all costs, and the idea of artificially odorizing oneself through the use of perfumes and the like is quite repulsive. The Huffington Post has an article about a woman who wears makeup for a month without washing it off.[115] The skin's ageing effect is quite alarming (see for yourself), but the only conclusion really pushed is to make sure you wash it off each evening. Such shortsightedness! No amount of creams and pastes and makeup will make you look better than you do already! Try and be more self-confident and recognize your own beauty. Unleash it! The paints do nothing but accentuate your lack of inner strength.

The mere notion that the body should be tampered with through the completely inessential utilization of cosmetic surgery is even more sickening. We are all born beautiful, and if we have lost that beauty through the constant sullying and scumming of our internal organs, then the only real way to regain it is to set upon the perhaps long and often arduous journey of cleansing it and restoring it to the health of our given birthright, which we have chosen unwittingly to trash.

Of course, awakening to Eden Fruitarianism can occur at any stage of one's life, and we will all have a history before arriving there. If it should so happen that yours included the piercings, the tattoos, and the regular daily appliance of skin and hair paints to enhance or cover up certain features you have deemed not good enough and have been unhappy with, then just accept this history. There is no point in giving yourself a hard time about things you were ignorant of; just learn from your mistakes. If the damage you have done is irreversible, just accept that you've erred, and commit yourself to not doing so again or at least to attempt to not do so again.

Chapter 8
Pet Peeves

I've written of how we need to change our attitudes toward species other than our own—how we need to stop seeing them as food, and start respecting them as individuals. Indeed I am guessing it is likely clear to most readers that animals have no desire to be imprisoned and murdered for their flesh and skins. It should be clear too that no animal would be happy to be confined and experimented upon, as is the plight of those suffering from vivisection. Equally, the whole concept of animal circuses and "zooillogical" gardens should be recognized as fundamentally exploitive of all animals involved.

As a quick side note, I just want to congratulate certain countries on recognizing the inappropriateness of animal abuse in entertainment. There is a slowly awakening global realization, and several countries are forbidding the use of animals in circuses[116] and even zoos![117]

What many will also tend, or choose, to be unaware of though is the exploitation of the animals right under their very noses. I am talking, of course, of the Rexes, Tabbies, Scruffies, Sooties, Snowies, Sheps, Gingers, Misties, Oscars, and Rovers among us.

From an Eden Fruitarian perspective, there is little to zero justification for continuing the practice of keeping companion animals, or pets as they are more colloquially referred to as. The way I see it, the vast majority of pets are only companions for as far as it suits their owners' needs and comfort. Even the whole concept of someone "owning" someone else I find rather sickening, and to be sure it is a legal "ownership" we are talking of here. There is a clear master and a clear servant, even if some pet owners may frequently joke otherwise.

Cats and dogs, probably making up the bulk of the companion animal market, are forcibly removed from their broods, litters, and families, exchanging hands rather like slaves once did between humans, and often by a business or individual who reaps financial reward through the transaction and never with full concern for the animals' own true needs, comforts, and desires. I have no doubt that, in many cases, both offspring and parents are left traumatized through the experience, if only for brief periods of time during transition. Of course, humans will make all manner of excuses and justifications that they fully believe they are somehow doing the animal a favor and offering it a good, warm, loving home, where the animal will be fed, pampered, appreciated, and otherwise well taken care of.

The truth, though, is that the keeping of animals is pretty much always exploitative of the animals in question. They are there because their keepers have chosen to have them there, and rarely because they themselves have consciously chosen their environment. Thus, the predominant reason for keeping pets is clearly a selfish one. Throughout their imprisoned lives, the vast majority of companion animals are left to spend hours alone. Dogs are often driven to temporary or even permanent insanity through boredom and loneliness, barking incessantly for hours on end at any untoward noises. The owners are often fully oblivious to the dog's failing mental welfare, as our canine friends have an innate capability for forgiveness and will generally immediately become overly happy and enthused on the return of an absent master who has been gone all day.

Honestly, can we truthfully say we deserve such loyalty, faithfulness, and forgiveness? Isn't this really just an example of how we exploit these animals for our own selfish reasons? With human companions we have to work out our relationships, knowing we may not be forgiven for whatever we do and that we cannot simply take without giving anything in return.

Cats generally have a fairer deal, as in many ways they are more independent than dogs and can be trusted to be alone outside for the bulk of the day, but they too are very much enslaved and ultimately degraded through their dependency on humans. It is humans who decide what, when, and where they can eat and, exercising our dominance over them, we frequently have them physically mutilated such that they are no longer able to procreate. Dogs we have even more control over, deciding when and where they get their free time, exercise, and play, and who and which other species they are allowed to interact with. We drag them around on leashes, only let them poo and pee when we feel the time and place are fitting, and angrily bark and snarl instructions at them when they do what comes naturally to them and disobey or behave in ways fitting of a dog.

Actually, I just wrote that cats can be trusted to be alone outside all day, but, well, that trust can only be extended toward their encounters with larger species, and many a small urban wildlife falls victim to the currently natural hunting instincts of domestic cats. But I intend to get to more on this later in my next chapter on NATURE.

The truth is, too, from a vegan perspective, cats and dogs are often fed diets consisting predominantly of the flesh of other animals. Now if you call yourself vegan and truly understand the guidelines of being a vegan, how can you possibly rationalize such actions?

How Many Animals to Feed Your Dog?

I read once, somewhere, in a vegan magazine maybe, probably back in the late 1980s, rough statistics someone had knocked together about how much an average dog of an average size might eat throughout the course of an average doggy lifetime. I wish I had cut out the article, as I believe its fundamental reasoning was pretty soundly logical and to-the-point. The facts were laid bare and the statistics showed that such an average canine could eat the equivalent of one cow, three sheep, three pigs, a hundred chickens, countless fish, and of course a fair bit of veggies and grains too, which are irrelevant to the current thought experiment. The numbers given here are purely my own fantasy, as I see no real point in calculating things more precisely. Whether it's one hundred chickens, twenty, or five hundred makes little difference; the truth remains the same, whatever the actual figures may be: if you think you are helping animals by keeping a dog and feeding it a traditional canned diet, then think again. By doing so you are contributing toward making the lives of many other individuals a living hell.

Of course, dogs can be fed vegan diets and are said to fare exceptionally well on such. In fact, I also recall reading some years ago that the world's oldest living dog at the time ate a vegan diet and ate much of its food raw too. Brambles was her name, and I read later she died in 2003, pushing twenty-eight.[118] (That would have made her well over 190 in dog years!)

Converting your dog to a vegan diet will naturally help to relieve the bloodshed on the planet and our over abusive nature toward animals, but considering not trading in further potential animals after the dog passes over will aid yet further.

Cats it seems have a more difficult time existing on a vegan diet, and I'm told success is only achievable via synthetic taurine (a protein they apparently cannot live well without). Although the statistics for what a cat might eat throughout an average life are likely somewhat less than the average dog, the figures would nevertheless still be comparable. The way I see it, keeping cats and veganism, particularly Eden Fruitarianism, are just not compatible. If you don't find yourself being forced into a compromise and buying the butchered flesh of other animals, then you'll still have the issue of the spaying or neutering and likely cause the consequential loss of birds and similar small wildlife from your back yard, which are partially your responsibility (by some estimates, over three billion birds are killed annually by cats![119]).

Unfortunately, like shoes, many people have cat and dog fetishes— even to the point where they claim to love their pets more than their husbands, wives, or own children—where they believe they are loving owners but in reality fail to see the full picture. It surely seems to me like something must be wrong there.

Of course, cats and dogs are not the only animals kept as pets; some fair far worse, like fish confined to small tanks for the amusement of guests and owners alike, and birds who are never allowed to stretch their wings imprisoned in cages, or if allowed out, often to have their wings forcibly, periodically clipped. Out of curiosity, I picked up some fish food flakes thing recently and looked at the ingredients, and even that wasn't vegan! It had powdered fish in it!

Then of course there are the hamster, mice, and other rodents people enslave for life, and, less commonly, lizards, snakes, and other reptiles too—none of who ever gets a truly fair deal.

If you think you are doing an animal a favor by rescuing it from a pound, then you are doing countless others an injustice. By doing this, you help perpetuate the collective thought that keeping companion animals is normal, and as soon as your animal is removed from the pound, it will be replaced. I'm afraid I'm with Spock on this one, and think that the needs of the many far outweigh the needs of the one, and as long as we keep on supporting this pet industry, the bloody pounds will continue to exist. It's a vicious circle that can only be brought to an end by going off on a tangent and washing our hands of it.

A Horsey Story

Just recently I ran into an acquaintance at our local market. She and her family keep horses as pets. Her husband was throwing snide comments around about meat eaters and their ilk, and how he wouldn't trust to have them on their land for fear of their horses being eaten. Now I understood he wasn't being totally realistic with his scenario, but for some reason it really rubbed me up the wrong way, and I immediately began questioning why it was they kept horses in the first place. Not liking my question, the husband left pretty much immediately, leaving his wife and me to discuss the topic in more detail.

Now, bearing in mind they are self-declared raw vegans, and quite outspoken at that, her justifications seemed pretty bizarre. To begin with, her major argument appeared to revolve around a supposed fact that, to be honest, I had never heard of nor consequently given any thought to. Apparently, she began telling me, horses are unique in that they, and they alone among species, have a gap between two different sets of teeth, and that therefore horses must have been created by God to be faithful companions to humans, as the gap was a perfect fit for the bit (a mouthpiece, typically made of metal, that is attached to a bridle and used to control a horse).[120] To me this response was comparable to the meat eater who defends meat eating by saying, "Animals must have been created by God for us to eat, otherwise they wouldn't taste so good." You can't really argue with that one.

She continued, stating that a horse's back was perfectly designed for humans to sit upon, and supposed that in the garden of Eden we would ride our trusty steeds at exhilarating speeds through the orchards to get much sought after adrenaline rushes and reach fruit three kilometers away. I'm guessing her version of Eden wouldn't be exciting enough for her without them. Frankly, I found her opinions at first to be rather vexing, in that they didn't really seem to make much sense to me. I countered by informing her that wild horses naturally don't mix with humans and will do their best to avoid them. In fact, the only way horses can be exploited as we do to them is through a lengthy, painful, humiliating breaking-in process, which she somehow managed to believe was irrelevant since that had happened generations ago.

Actually, later on when I got home, I felt a need to get online and verify her statement about horses being the only animals with that gap. I doubted its authenticity and, after wielding my trusty Google, soon discovered that cows and most other ruminants have the gap that is called the "interdental gap." In fact, zebras and donkeys have them too. A little further research showed me—the percentages vary depending on which website I looked at—that somewhere between 50% to 80% of horses (as opposed to, from what I can tell, 0% of cattle!) actually do grow teeth in the interdental gap. It's something called "wolf teeth," and if they are allowed to grow, it would make the usage of a "bit" extremely uncomfortable, even painful. Thus equine dentists are routinely called in to remove them.

I am guessing that, as this lady and her family have six or seven horses, they would likely have been aware of this fact and thus, in justification of their animal dominance outlook, deliberately did not disclose the full truth.

Discussing this further with Květa, she and I reflected on the supposition that a horse's back was designed for humans to sit on, and Květa pointed out if that were true, people wouldn't need to create and use saddles to make the ride more comfortable. So true.

I find this kind of attitude actually quite typical of pet owners. People have attachments toward their pets, whatever animal species they may be, and God forbid should anyone dare challenge that attachment.

What I find odd, though, is rather like regards to diet, where people consider some species fair game and others off-limits; with no particular logic behind their conclusions, people have a tendency to view other species as off-limits in terms of companion animal options too. Take the discussion about the horses. I made a quite logical comparison toward keeping chimpanzees as pets, as opposed to the keeping of horses. I proposed a similar scenario, where a theoretical chimpanzee family had been in captivity for a given number of generations and had thus lost all its wildness and would even seek out the company of other humans, preferring time spent with them, their company, affection, and food shared with them, over even the eventual sometimes given available freedom of the outdoors. I asked if she would consider that having such an animal as a companion would be acceptable, but she refused to go there and, clearly a little offended, informed me that "horses are unique" and that such a comparison was invalid and that the gap was categorical proof. I countered that chimpanzees were unique too; indeed, all animal species are unique. Shucks, even every individual member of each species is unique.

In retrospect, I think I should have chosen zebras or cows as my comparison animals (especially had I been aware that these animals shared the supposedly unique-to-horses interdental gap!).

And one could well-imagine that these animals also, over time, could have their wills broken—wild and untamed spirits subdued—and become in servitude to humans. I guess I just wasn't thinking fast enough.

Anyhow, there was a lesson in there for me too, as I came away feeling agitated and unsettled, and know it was because my whole attitude had been wrong. Getting wired and fired up about issues that ultimately were beyond my control was not productive for anyone.

Seeing the Full Picture

Likely there are many who have bought animals in the past, who would find this whole chapter offensive to them, believing that I'm not seeing the full picture, when to be honest, I feel I see it far more clearly than most.

I guess many might counter that some animals have been domesticated over centuries, so this somehow justifies the perpetuation of such dominance, but personally I don't fathom how this should mean it is right to continue such perverse customs. Slavery was once widely practiced in the past, but is today just as widely condemned.

As for animals choosing to live closely to us—I'm anthropomorphizing again—as a species I'm guessing they did not expect to have to give up their freedom, and the world was a very different place back when it all began. Humans themselves were living in low-tech, close-to-nature ways, not in cities with motorcars and central heating and canned pet food.

I like to believe humans will evolve their consciousness, realizing that pets are not a natural addition to one's family, and that the pet industry will eventually see its demise. The only animals who have complete freedom and autonomy are wild animals.

I understand that the occasional companionship of a nonhuman animal is said to have therapeutic value, but so can that of a human, and it is because we have isolated the aged and disabled, and no longer live in small close knit tribal communities, that the need for such company arises. Even better company would be another loving sympathetic human companion. Of course, having a human or animal friend is not exactly comparable, I see that, but I refer again to the "everlasting" ability of a dog to "forgive" (no matter what you have done!) and to remain loyal. That may be true, but do we really deserve that? What do we learn if we are continuously forgiven for mistreating and abusing others?

Robot Pets

As a side note, and as a possible future relief for pet people, I hear the Japanese are making progress in creating robot pets that may one day replace our fellow sentient beings as faithful domestic companions.[121] This I see as far from ideal, but at least it could mean animals would cease from being exploited for the purpose. This would be somewhat like synthetic meat to the newbie vegetarians, or soy cheese to the lasagna-craving vegan!

The fact that the majority of pets are spayed or neutered should be a deal breaker for many with budding awareness. The sexual mutilation often has more to do with maintaining our control over their lives than any real concern about uncontrolled cross breeding, and is another predicament that we have gotten into by keeping animals as pets. Spaying and neutering animals is committing gross acts of violence against them, a clear question of animal rights. Spaying and neutering denies animals their right to mate, to be parents, and to have families of their own. If you somehow think the end justifies the means, then tell me how complacent you would be to be forcibly neutered, or how right you think it would be to forcibly mass-neuter humans in overpopulated areas of the Earth?

Whilst the present attitude toward pets continues, dog homes will never run out of animals. Only a small percentage of abandoned and ill-treated animals are ever rescued, and here's an interesting statistic: over one thousand homeless dogs are slaughtered weekly by the UK RSPCA alone! Likely a fair percentage of them will themselves end up as pet food.[122] I quote:

Many of these animals [that end up in pet feed] died after being medicated for health problems that contributed to their deaths, and not all drugs are neutralized during the rendering process. Meat and bone meal can contain antibiotics, steroids, and even the sodium pentobarbital used to kill pets at shelters. By definition, a lot of the animals that ended up in the rendering vat had something wrong with them...

This is the food, which is considered unfit for human consumption, that is often unquestioningly fed to companion animals.

There is no compassionate way forward that includes enslaving animals, unless these animals are completely autonomous, free to come and go as they choose, to be with their own kinds if they wish, and to do all the things wild animals do.

The best way that children can learn to respect, love, and value nonhuman animals is to see them and learn about them in their free and natural state. Having pets does not teach respect for animals. If that were true, then all pet owners would be vegan!

I understand every situation is different, and there are cases where animals may choose us too. I've been fortunate enough to experience this on more than one occasion, and can vouch that there is a great joy to be had in gaining the trust of a wild animal. In such situations, I believe we should feel privileged and blessed when a wild animal chooses to befriend us, and that we should not take advantage of its trust.

The term, "pet," is disputed by some concerning their own choice(s) of companion animals. My colleagues with the horses would likely never call them pets, but, well, it's just semantics and by definition a pet is a domesticated animal kept for companionship and/or amusement, and I certainly see this as the predominant reason that they are keeping horses. If the horses never sought out their company, refused to let them approach or mount and ride them, shied away from them, and kept to themselves at the back of their land, then I seriously doubt they would ever consider reinvesting in new ones. Actually, I've since witnessed how their horses are kept mostly penned-in or tied to stakes for long periods of time, and realized just how much denial there was in the "horses running freely scenario" she had described to me.

Animal Lovers

It's an unfortunate reality that many pet owners mistakenly consider themselves to be animal lovers. Indeed, I recall a survey where participants were questioned as to whether they believed in animal rights or not. Quite a high percentage answered yes. It turned out, though, that they had believed the question related purely to dogs and cats. For them, being an animal lover means loving their shiatsu poodle and thinking nothing of eating a heart (attack)y egg and bacon breakfast.

I occasionally feel the urge to comment on those Facebook images of a dog or cat being mistreated with the accompanying text that animal abusers need to be severely punished. You'd be surprised how emotional some commenters can get, crying out for the death penalty for those who hurt animals.[123] Not too surprisingly, whenever it is pointed out that the logical conclusion of this is to become vegan, cognitive dissonance steps in, and the most irrational answers are given that frequently make little to no sense.

Then there is the issue of toxins in animal fur. Toxoplasma gondii is one particular parasite found on house cats that is easily transmitted to humans and can cause serious harm to unborn fetuses. Apparently it takes up residence inside the human brain and some statistics estimate as many as 50% of us are infected with it! Okay, it's said to be relatively benign, but studies have suggested that there's a correlation between this parasite and several different personality traits.[124]

When broaching the subject of keeping pets on my blog in the past, it was suggested by more than one reader that I was perhaps being a little too "blanket" about the overall situation of pet keeping—that things aren't always as black-and-white as I make them out to be, that there can be mutual benefits for both the companion animal and its owner, and that other animals need not consequently be slaughtered nor harmed in the process.

Okay, so let's examine such a hypothetical situation. Firstly, it is pretty clear to me that those who defend pet keeping in such a manner are likely doing so because they themselves are attached, in some fashion, to the idea of animal keeping. Secondly, although I don't altogether gainsay the possibility of such a scenario, I just see it as being highly improbable. There are always issues and situations that arise making it profoundly clear that we are still most assuredly talking of a master/servant relationship. I think, for example, we can safely rule out ever fitting city-kept dogs into such a postulated synopsis. They rarely get the free range they require and desire, and there is often the additional problem of their discordant barking disturbing what little peace there is to be had in urban settings. I'm sure most city dwellers and many country dwellers have had their sleep disrupted with the incessant yelping of bored nighttime canines, such a contrast to the gentle chorus of frogs and crickets.

Even out in the country, where a dog gets to live with several of their own kind, is given more or less free and unfettered range, and is theoretically still fed a vegan diet, there are a multitude of other factors to consider. Least of these factors is the high likelihood that the ingrained pack hunting mentality and instinct will resurface, causing terror for not only local wildlife but also for those of us who enjoy a pleasant country stroll. I've lost count of the amount of times free-roaming dogs have caused me momentary fear and compelled me to change my trajectory.

And if they are not sexually mutilated and allowed to freely mingle with their own un-mutilated kind, there is the concern of inevitable offspring, perpetuating the problem ad infinitum. Most people are unaware that roughly only one in ten to twelve dogs and cats ever find a home.[125] Many of the rest end up being sent to rendering stations where their carcasses are processed into, among other things, pet food. In the US alone nearly ten thousand pounds of animals are slaughtered daily![126]

In a Nutshell

I recall witnessing a vendor sitting outside of a Buddhist monastery in Thailand. Beside him he had a cage full of some kind of small squirrel-like rodents. A likely devout Buddhist exited the temple grounds and approached this man. They exchanged a few words and the Buddhist handed him some money. The vendor reached into the cage and took out one of the small animals and let it loose on the ground. The Buddhist bowed and walked away. He'd done his good deed for the day and freed an innocent animal, thus in the process (and I'm sure this was his logic) lessened his karmic burden. I stood by to see what would happen next. The animal, bewildered by its newly presented freedom, sat there confused-looking not knowing where to turn. The vendor calmly grabbed it and, after returning it to the cage, sat and waited for the next customer.

Now, I'm hoping it is clear to most reading this that the Buddhist really did nothing to aid the plight of the imprisoned animal. In fact, if anything, he was the partial cause of the animal being imprisoned to begin with. Without him and others similarly shortsighted, the vendor would gain no financial benefit from capturing those animals and sitting there outside the monastery.

This is how I see the pet industry being kept afloat—by people who, although they basically mean well, are just too shortsighted to see the bigger picture. They may kid themselves into thinking they are animal lovers and are saving the life of one animal, but the reality goes much deeper. They are financing a trade that is riddled with animal abuse.

As for the argument from the owner who claims to love their pet more than anyone or anything else—if anything, this should make one question society as it is, and one's warped idea of love. I read a story recently about a Dutch woman who regularly visited a zoo in Rotterdam.[127] She went specifically to observe a particular gorilla named Bokito, believing she and the gorilla had a special connection. One day, the gorilla escaped its enclosure and, with great deliberation, headed straight for the woman, whom he grabbed, mauled, and dragged around like a rag doll. Apparently her constant staring at him was winding the fellow to the point where he could no longer stand it. Despite the water separating them, and the fact that gorillas don't swim, in a fit of rage, he just went for her.

I honestly think these so-called animal lovers often don't really have a clue. I know the gorilla story isn't really related to the pet issue, although the concept of keeping any animal in captivity is pretty repulsive. I stand by what I say and believe that as long as this pet mentality remains, the accompanying pounds (animal concentration camps) will forever be kept open.

I am not trying to point the finger at anyone particularly, and for those who have already entered into the commitment of owning and caring for a companion animal, I don't wish for it to sound like my recommendation is for you to ditch the animal (even though, yes, I strongly believe it most likely was a mistake getting it in the first place). But if that damage is done, it is our moral obligation to come to terms with it as best we are able. I only hope that some deeper thought on the subject might lead one to think twice before allowing oneself to make similar such future choices.

Respecting the Wild

Should we be privileged enough to be in regular contact with genuinely wild animal species, it is important for any friendship with them to not revolve around them losing their independence.

For an example, the previous tenant of the land we are now joyfully living on began the habit of feeding the local kookaburras butchered diced flesh. To be sure, the birds soon learned of the regular free meal thrown onto the rooftop of a utility shed and began just hanging around in anticipation.

There is no doubt that the fellow meant well, but he did not realize that by befriending the kookaburras in this way, he was effectively disabling them, making them dependent and lazy. They no longer spent time looking for their own food but instead would just wait around to be fed. When we moved in, they continued expecting the free meals and would call out angrily for us to feed them. For a while, the old tenant continued to come once or twice a week to throw chopped meat onto the roof. It took us a while to realize what he was doing and, putting two and two together, asked him politely to please stop.

The kookaburras went through a period of transition, where for a while they would hang around expecting once more to be fed, calling out whenever we ventured in the direction of the shed. It appeared too that their feathers looked slightly unkempt, as if they were not cleaning themselves properly.

After a while, they began to get used to searching for their own food again. They still live here and wake us up each morning with their melodious laughter, but they've now regained their stolen independence and once more find their own meals (Their feathers definitely appear to be glossier now too!). This is probably fortunate for us since occasional venomous snakes are around, and the local kookaburra population helps keep them at bay.

In 2010, we were living in a house in town. We would often be out in the small garden we had, where we grew, among other things, tomatoes, cucumbers, papayas, and passion fruit. A small bird would often be hanging around, a native willy wagtail, a regular guest star in suburban gardens here. I used to talk to him as I gardened and, one day, he bravely flew onto my head, touched it briefly, and flew off again. Over the following week or so, he did this more and more frequently. He just flew up and touched me, on my arm or leg or head. It did not take long before one day he settled firmly onto my arm. I looked at him, smiled, and talked to him. He flew off, but from then on he would regularly fly and land on me as I gardened.

Some days later, he also began showing such familiarity with Květa and, much to her joy too, began landing on her while in the garden. He would even sit on the windowsill of the house and call out for us to come outside. After he had built his home and became the proud father of three small offspring, he called me to see the nest in the back garden mango tree. Beautiful precision work and three very cute babies.

We never fed the bird, so I'm supposing the relationship was purely one of friendship. This went on for about a year, and we took quite a few photos of it with us. Eventually, we are guessing, the neighbor's cat might have gotten him, as he completely disappeared from the garden.

Pet Poop

Then there is the issue of the toxicity of pet poop. Unlike many other species feces, cats' and dogs' are particularly toxic. You never hear of people putting dog shit on their roses. The shit of flesh eaters is always more toxic than that of herbivorous animals, which is why it is not recommended for the conventional dieter to recycle their own waste in their gardens. The story is quite different for fruitarians.

Summary

To summarize why we should reconsider support of the pet trade:

1. It keeps those animal species involved enslaved.
2. It necessitates the murder of many other animals.
3. It causes the murder of many other dogs and cats.
4. It causes and supports bloody sex mutilation.
5. It causes animals to be driven insane from boredom and loneliness and separation.
6. It causes noise pollution (bored dog barking).
7. It makes us think we are animal lovers when we are not. (Animal lovers don't eat animals!)
8. It restricts our own personal freedom, as such animals are always dependent on us.

All companion and domesticated animals have dampened spirits. Compare a wild brumby[128] (feral horses of Australia) to a domesticated horse. While the wild brumbies roam powerfully free, with terrains spanning up to hundreds of kilometers, their subdued shod counterparts stand listlessly forgotten in fenced paddocks.

Their sparks are gone. Their spirits blackened. Vitalities just deranged fractions of their potential glories. Compare the house dog to the wolf. Can you not see the injustice? Their liberties must be reinstated!

Chapter 9
Religion And Spirituality

You may have jumped to the conclusion that Eden Fruitarianism is in some way connected to a religion, especially considering Eden is such a historically biblical concept. This is far from the truth though. Eden Fruitarianism does not identify with any specific religion, cult or sect, orthodox or otherwise. Indeed, whether or not the biblical Eden ever truly existed is really irrelevant to the issue.

You may also have initially picked up this book thinking it was about diet, but that too is a false assumption. The crux of Eden Fruitarianism is really about recognizing sentience, empathizing with it, and doing the right thing by all parties concerned.

Pushing aside very personal beliefs in deities and fairy godmothers, if I am not mistaken, this also just so happens to be the neglected crux of all religious teachings.

The Golden Rule

Yes! The profound essence of all religions is really about doing no harm, treating others with the respect we ourselves expect or fundamentally deserve, and generally loving and helping others when and where we are able. This is what scholars have for centuries called the "Golden Rule."[129]

Buddhism, for example, has what it calls the five precepts. The first of which clearly states that one should undertake the precept to refrain from harming all living creatures, both humans and other life forms alike. By inference, it seems that the ultimate harm is the taking of life! I'm sure there are Buddhists who understand and act fittingly upon this tenet, but to be honest, I've yet to meet any who go out of their way to put this into practice, and I've even read that even the Dalai Lama regularly eats the flesh from other beings.

Christianity, in addition to Jesus's teachings of respect for others, has the very clear sixth commandment given in the Old Testament "thou shall not kill" (Exodus 20:13; Deuteronomy 5:17). Matthew 7:12 says it pretty clearly too with, "So in everything, do to others what you would have them do to you, for this sums up the Law and the Prophets."

Speciesist scholars addicted to the scorched flesh of butchered animals have tried to distort the true meaning of the sixth commandment and would have us believe that it refers purely to the murder of another human being, but I would seriously doubt that any such exception was ever implied or intended. Some have been even more liberal with their interpretation and narrowed the definition down to "human being who is on your side." History is riddled with wars fought in blasphemous uses of God's name.

Islam, supposedly the fastest-growing religion on the planet,[130] quotes several times in its Koran that one should show respect for animals: "but to hunt is forbidden you, so long as ye are on the pilgrimage. Be mindful of your duty to Allah, unto Whom you will all be gathered" (Koran, surah 5, verse 96).

This teaching states that in Mecca, the birthplace of Mohammed, no creature can be slaughtered and that perfect harmony should exist between all living beings, which is odd, as I can't quite imagine that all the restaurants there are vegan.

"Whoever is kind to the creatures of God is kind to himself" (the Prophet Mohammed).

So it seems Mohammed was not always kind to himself, as he also partook in feasting on the flesh of others.

"There is not an animal on the earth, nor a flying creature flying on two wings, but they are all peoples like unto you" (Koran, surah 6, verse 38).

Ahimsa is an East Indian term relating to the principal of non-violence toward all life; it is an important tenet of both Hinduism and Jainism, where animals are seen as sentient beings worthy of human love and protection. Killing animals for food or any other reason is supposedly completely unthinkable to an adherent of Ahimsa, although I suspect that, as with other religions, there are disciples who will ignore or regularly turn a blind eye toward this tenet. Violence toward animals is said to negatively affect a person's destiny and fate. The concept of karma suggests one reaps what one sows, and adherents of Ahimsa believe the violence and misery experienced by a murdered animal will be metaphysically transferred to the flesh-eater during their own lifetime.

The Bahai Faith teaches that respect for animals is just and even encourages its followers to become vegetarian. Some will even choose to cut out all animal products and eat vegan. Admittedly, not all followers have the insight enough to equate respect for animals with diet at all, so I'm guessing fully committed vegetarians in the faith are most likely the minority.

Furthermore, Abdu'l-Bahá said, "Fruits and grains [will be the foods of the future]. The time will come when meat will no longer be eaten."

The next time those Jehovah Witnesses come knocking on the door, take a look at one of their small colorful brochures. They often have one baring a picture of an idyllic setting with colorful fruit baskets and lions lying peacefully next to lambs. This they see as a future paradise, clearly portraying the Isaiah prophecies,[131] of a time when carnivorism will stop, even as a behavior between other species. To me, it makes perfect sense; for in order to bring about such a setting, we will all need to tow the line and change our eating habits. Apparently this is a fact that not one of those young Jehovah Witnesses I've spoken with has ever understood or appreciated. It's like they expect the changes to happen without changing anything!

I'm quite certain a few Google searches would show similar thoughts expressed by other religions, quotes that are generally swept under the carpet by leaders, scholars, and followers of each religion. The Pope, head of the Catholic faith, is a regular flesh eater, and so is in this respect similar to the Dalai Lama, devouring gluttonously all creatures great and small landing burnt on his plate.

It appears clear to me, I repeat myself, that the core of all religions is non-violence; and the most non-violent of all food (thus the pinnacle of religious food) is unquestionably fruit. It has the potential to be obtained with no exploitation, killing, or theft from animals or plants. No unjust imprisonment, branding, castration, torture, ear punching, mutilation tags of voiceless innocent animals. No acre upon acre of annual mono-agriculture, which is reliant on heavy industry tractors and harvesters, resulting in the loss of many small animals' lives and the loss of habitats for creatures of all sizes. The statistics for annual agricultural accidents and deaths to humans alone is quite shocking, not to mention the millions of mammals and other creatures sliced by combines, harvesters, etc.

Corruption

Unfortunately, as organizations, there are no common religions that haven't lost sight of these truths. Followers instead lose themselves in the quagmire of dogma grown around each religion, and often erroneously believe that they alone are in possession of the truth and that all other religions are inferior to their own. Unless you share their beliefs, you are in the wrong, and are possibly running the risk of experiencing eternities of brimstone and treacle. Instead of unifying humanity and bringing more harmony and compassion into the world, religion has regularly resulted in the opposite, blatantly encouraging discord and disharmony, and generally run with an almost militaristic hierarchy.

I believe that as spiritual teachings become slowly institutionalized into religions, corruption is a consequential inevitability. They become businesses and, as such, the focus shifts from the essence of the original teachings toward one more of power and financial revenue raising gain. Of course, this is often done cleverly so that the sheeple who flock to their congregations often have no idea of the corruption inherent within their particular chosen brand of religion. Some will even go to the extreme of murdering others who criticize their faith; indeed, there are probably people who will dismiss this whole book and all its ideas as nonsense, which is based purely on the fact that they feel I may have slighted their religion.

Of course, I am not at all denying inherent truths to be found within religious teachings, nor suggesting one should not believe in that which is beyond the physical. I myself, though not altogether comfortable with the word "God," have unbounded faith in the Divine, the great white spirit, call it what you may. I believe there is an overall fairness in all that exists, and that beyond the physical we are all eternal beings. I just do not believe any one religion has a monopoly on truth. It is somewhat like the "Golden Buddha" story revisited, where the essence of religion is pure spiritual gold. All the rest is mud. The elitism, the gay discrimination, gender oppression, genital mutilation, abusive dogma, the Christmas turkeys, the halal, the ghee, the threat of eternal brimstone, and treacle are all mud, and not even the glorious sort one can contentedly wallow in!

Karma

Our planet is full of laws, some of them contradictory, many of them ridiculous, and many of them meant only for the masses and not for the ruling elite. Some people have found ways to successfully circumnavigate certain laws and disobey them regularly. However, beyond the loopholed laws of humans, there exists a simple yet profound spiritual law of cause and effect. It goes by the title of "karma," and one cannot hide from it.

The way I see it, karma is a true universal law. It is a kind of cosmic accounting system ensuring that, despite often appearances to the contrary, there is fairness to everything. For every cause there is an effect. For every bad deed committed, negative karma is incurred, thus increasing suffering; and for every good deed, the karmic burden is lessened. I believe it is because there are so many bad deeds committed that there is so much suffering within this world.

Who gets to decide what's classified as good or bad? Some may deny that such a distinction even exists, but deep down our subconscious, our inner beings, our souls, always know. Certainly we are generally, easily, readily aware when someone commits a bad deed against us; we just have to learn to be more consciously aware when we do likewise against others. We also need to expand our definition of "others" far beyond the limitation of our own race, gender, ethnicity, and species. For example, should someone punch us for no reason, we would judge this as a bad act. If a dog bites us, we may think the dog is behaving badly, especially if unprovoked. Even if the dog is hungry and bites with the intention of eating us, we would surely protest. Should we boot the dog for no reason, we are the ones behaving badly. It does not matter that the dog is not human. Consider now whether we should force a pig into a life of misery and suffering, to finally murder and butcher its cadaver. Should this be considered acceptable behavior because the intention was to eat the fellow all along? Come on, use your imagination. Use your empathy!

If you truly are capable of doing so and can keep doing so, then eventually you too will come to the only logical conclusion that, as I have repeated on numerous occasions in the past, fruit is the only food that is truly given, both freely and generously from the plant, which willingly bequeaths it. Every other food incurs some kind of karmic debt.

A bad deed can be anything from murder, greediness, being deceitful to others or oneself (denial), not being affectionate enough, or deliberately turning a blind eye to the plight of a calf forcibly removed from their mother, robbed of its mother's affection and milk, and slaughtered for its veal. The combinations are endless. Callous uprooting of trees. What goes around comes around.

Reincarnation

But how can it be true that there is prevalent fairness throughout existence when there is so much about this world that on the surface seems so frankly unjust? How can it be fair for one person to be born into a loving family where all their needs and comforts are met, physically, emotionally, mentally, and spiritually, where food is ever present and a safe environment provided; and yet another, in stark contrast, is born into a war-torn country, surrounded by poverty, or to drug abusing parents who fail to provide even the most basic of needs, who may even consciously abuse their child, physically by neglect or in other ways?

Such scenarios are sadly all too common and can easily make one question whether this fairness I propose actually exists, or whether it is perhaps just a figment of my overzealous imagination. I believe reincarnation sheds light on this apparent dilemma.

It would be so much easier if when we pressed down the wrong button, we received an immediate reprimand. Even the slowest of us would soon learn that there are consequences to pressing such buttons. Unfortunately though, the universe does not work this way. There is no time limit on the effects of deeds.

Some responses are clear to some of us as to why we suffer the slings and arrows of certain outrageous misfortunes. Why we endure the regular bout of "the cold," for example, may be quite blatantly obvious when seen from the clarity of eating raw. However, when trying to explain why you just cut your finger, stubbed and broke your toe, twisted your ankle, or got knocked down by a random cyclist, there appears to be a randomness involved beyond any immediately recognizable cause and effect. Even with a more heightened sense of awareness things may happen that result in one saying, "Why me, why God, why?"

The reason may not at all be apparent, but I believe whatever untoward should bechance us, there is always, always a karmic explanation, quite possibly not even extending from this current lifetime. On the surface this can be confusing, because generally we have no recall of our past misdeeds. However, deep down, I believe there is an awareness below one's waking conscience, and if we can only acknowledge this fairness and understand that we cannot ever truly hide from our delinquent shenanigans, this can shape our present and ultimately the future of all sentience.

However, with this in mind, it is important we don't fall into the trap of thinking, "Oh, it's okay; what they experience is just karma, thus all is deserved." If you are part of their suffering, you are yourself incurring karma, whether or not you are in some intricately complex way also helping the universe repay another's karmic debt. We cannot justify injustice with the belief that we are helping some system we will never really understand fully. Because we won't! Let the universe reap its own justice. The way to quickly settle karmic scores lies through understanding, acceptance, forgiveness, and compassion; through yearning for peace with every cell of our bodies and thus becoming a magnet for it.

Divine Faith

All we really need to be successful with Eden Fruitarianism is faith in the divine and the understanding that harm to a part, is harm to the whole. Without such understanding, we can soon become lost in fear, doubt, and consequently fail.

Of course, I may be wrong about this whole chapter. There may be no such thing as this karma I talk of. There may be no such thing as reincarnation. There may be no such thing as the soul. The Muslims may indeed be in sole possession of the truth. Or maybe the Eckists are. Or the protestants. Or the Moonies. Or the atheists. Perhaps it is all just inexplicable random atoms obeying physical laws we will never understand and there is no right or wrong. Perhaps Douglas Adams hit the nail on the head when he humorously suggested the whole universe was sneezed out of the mouth of a giant dragon and that we should fear the coming of the great white handkerchief.

The truth is that little of the universe makes any real sense, so we should all be forgiven for the beliefs and faiths we hold. Not wanting to be limited by the finite, the concept I have outlined above makes the most sense to me and thus it is where I chose to let my faith lie.

Circle of Compassion

Regardless of how big or small it may be, each and every one of us has a circle of compassion. It's really the magnet that keeps sentience on the world sentient. As the saying goes, "Love makes the world go around!" If the reverse were suddenly true and none of us had a circle of compassion, sentience would soon trickle to a halt. This is about as likely to happen as it is that the world's gravitational pull would be suddenly switched off.

I find it difficult to imagine that there may be any individual being alive whose circle is limited to purely themselves. Even the most cold-blooded of creatures, such as the crocodile, has a circle of compassion encompassing more than just its own unique being. If this were not the case, then nothing would prevent it from eating its own eggs or newborn hatchlings. Instead the mother croc will gently carry its offspring to a place of safety and some will even care for and protect their young for up to three months until they are grown enough to care for themselves. Of course, crocodiles are hardly brimming, bubbling, and overflowing with love. Far from it, their circles of compassion barely extend beyond their known family within immediate territorial range, but this still doesn't stop love from being the predominant driving force keeping the species continuing.

Lions, the degenerates of the jungle, will sacrifice themselves to protect their cubs, and have circles of compassion often extending to encompass the small tribes they live in. Some have even been known to adopt the young of other species. I recall one particular case from the beginning of the twenty-first century, where a lioness even adopted an oryx,[132] a species of antelope native to Africa! Then there is the case of Little Tyke,[133] another lioness, who refused to eat meat pointblank!

A circle of compassion is basically a way of describing the sphere of sentience that we care for. Even for some humans it may be just immediate family members and friends, and indifference toward the fate of anyone outside of the circle. It's likely that some of this group might include individuals not of our own species. Their cat(s) and/or dog(s) for example. Others might extend their circle to include all members of their tribe, village, city, country, or members of their own ethnicity, religion, culture, or supporters of their football team. Many feel compassion for every other human being and are saddened by any human suffering, seeing us all as human and worthy of respect and dignity.

Some—currently a minority—consciously choose to widen their circles yet further to embrace other species beyond the narrow confines of the canines and felines, and wish to avoid intentional harm to most other land animals: mammals, birds, and reptiles. From there, it is a small step to experience empathy for creatures of the oceans and waterways too and to wish them all no harm. These people understand there is no valid logical reason to exclude such creatures from their circles, especially not one based purely upon difference of appearance or intellectual capacities; instead they see no reason why they should be treated any less respectfully and not be honored.

Some circles have grown to include specific plant life too. This may mean caring for a particular prized potted plant that's both caressed and softly spoken to daily. Or it could be a much-loved tree at the bottom of the garden, appreciated for the shade it provides and its intricate branches perfectly suited for climbing, or simply the delicious mulberries it gives generously each year.

Then there are those who concede that the humble carrot, in its own silent way, may be none too agreeing with being uprooted and munched on. Such a person needs no scientific proof of this, just as he or she needs no scientific proof that injustice is inflicted on a goldfish who's confined to a small bowl, or that a canary suffers from its improper imprisonment, or that a dog is aggrieved from a sudden sharp smacking, and especially the realization that fruit is a viable solution to the live-and-let-die paradigm. Even if there is uncertainty regarding the experience of the likes of a beetroot, there can certainly be no harm in giving such life the benefit of the doubt and letting it live blessedly.

The good thing about the circle of compassion is its lack of boundaries. One can forever keep extending the outer limits. Looking beyond the mistreatment and abuse of animals inherent throughout most of society as we know it, we can see the destructive effect our presence has on the environment and the rest of nature, and do our darnedest to try to minimize it.

In defense of keeping the circle tight, people will often argue that you can't possibly ever encompass everything and thus it's pointless even trying. But this argument is so irrelevant. Even if there are compromises and things still "get hurt," there appears to be no reason for us to not make efforts and grow through doing so. It does not make us hypocrites if we acknowledge that there are times when our presence is unavoidably linked with the hurt of others. As we grow, we can learn to avoid such situations.

Tikkun Olam

I came across an interesting concept recently—something known as Tikkun Olam,[134] which is a Hebrew phrase that can be translated as, "healing and restoring the world." It suggests that humanity shares responsibility (with the Divine creator) to heal, repair, and transform the world. This is somewhat connected to the biblical fall from grace, and implies things aren't quite as they should be and so we as humans are partially responsible, if not for the fall, then at least for restoring the splendor of the pre-fall epoch.

The Long and the Short of It

Hopefully, these examples can demonstrate that Eden Fruitarians are liberationists and not somehow religious fundamentalists. Really it matters little what your own personal beliefs are concerning any divine deity, guardian angels, or potential afterlife, or if you should even choose to believe so or not. Eden Fruitarianism is essentially not about religious beliefs but about an understanding of the golden rule and an awareness of one's circle of compassion and the very real need to expand it. The afterlife that should be our highest concern and recognized as clearly influenceable is the world we leave our children.

Eden Fruitarianism is not about forcing a personal view onto anyone. It is about open-mindedly recognizing consciousness and questioning any variety of exploitation or violence toward any form of sentience. To compare it to religion is to fail to grasp the difference between faith and a moral argument. It's the equivalent of comparing human rights to a belief in the love goddess Ishtar.

Actually, I was pleased to read this morning that Pope Francis, unlike his conservative predecessors, stated that salvation is there for all regardless of religious faith—atheists too—and that it's obtained through good behavior, not through belief in Christ (as has been the prior Christian axiom).[135]

Section Three

THE VISION

Non-violence leads to the highest ethics,
which is the goal of all evolution.
Until we stop harming
all other living beings,
we are still savages.

- Thomas Edison

Chapter 10
Nature

Up until this point, I believe there is a rudimentary truth surrounding every issue raised so far, and anyone serious enough, who can put aside old prejudices and societal brainwashing and give more thought to these things, will understand there is a very real need for change. For sure, many will not find themselves prepared enough to embrace every notion I've aired, but the essence of each chapter's message is, I believe, indisputable. (Go ahead, feel free to try!)

This chapter introduces a future vision, a scenario that I suppose will likely, currently, only be embraced by a few. It is, I believe, a vision of positive change, and whether or not you are willing to share it with me does not deter me from pursuing it. Be prepared for something different.

The Food Chain

Even those who reside in cities with little contact with nature's wilderness will surely be aware, through the miracle of TV and the likes of David Attenborough, that life on Earth is rarely anything but peaceful.

Turn on any nature show, and the dog-eat-dog world we live in will be overwhelmingly apparent. Some of us feel discomfort when the crocodile's jaws clamp themselves onto the rump of a passing wilder-beast, yet it's a rare individual who will actually question whether or not this live and let die path is the only possible scenario for this world. Indeed, most people would argue that in order for any life to continue its existence, other life must inevitably be sacrificed—that it's just a sad fact of life, whether it be through the flesh eating of another animal or the untimely death of plant life, violence of some form is unavoidable.

Challenging the Paradigm

Have you ever paused to consider whether this really need be the case; is there perhaps some other way for life to continue? I have at times broached this issue with others, so I'm already aware that there will be a certain percentage of the population who will be shocked to hear me question the validity of the prevailing balance of nature. Some may immediately jump to the conclusion that something must surely be wrong with me to even question whether or not things need to be the way they are, perhaps even concluding that I am not only unrealistic in questioning it but also in some fashion proposing some form of disequilibrium or mass extermination of certain species.

However, for me, everything I envision throughout this book is well within the realms of the ultimately achievable. I am certainly not proposing any physical intervention or eradication of species I deem to be unruly, far from it. I am totally aware of how things are and am not the least bit in denial that live and let die is the current way of the world. However, within the expansiveness of an apparently infinite universe, I am convinced "balances of nature" can be achieved by less violent means, and that despite how the situation currently is, things can and will change. Change too, is an inevitable fact of life.

It seems to me that if one is forced to take the life of another in order to continue living, then such an existence is a very cruel, unloving, inconsiderate, malevolent, uncaring, contemptible, ungodly way of being. I suspect things may not always have been this way, and that at some point in the likely distant past there may have been some kind of, for want of a better term, "biblical fall from grace," rendering the world on its current path of chaotic disrepair.

However, whether or not such a fall ever took place is not really the issue in question. The way I see things is quite logical and can be viewed and understood by anyone prepared to invest some moments of contemplation to the following:

If someone attempts to harm me, with no apparent cause or provocation, I would naturally protest and attempt to safeguard myself against injury. Whether it be by fleeing the scene, attempting to appease said person through words, or through otherwise defending myself in some fashion is irrelevant. Regardless of whether such an onslaught should succeed in harming me, I would likely rule such an act as being unjust. If someone does likewise with you, I would undoubtedly reach the same conclusion of judgment. I find it quite probable that you too would reach the same conclusion given such hypothetical circumstances. No one deserves to be attacked without cause.

Now, if I am walking through the forest, minding my own business, and a bear happens to cross my path and attempts to attack me, I would still pursue the same course of action (short, perhaps, of trying to verbally reason with the old chap) and do my utmost to avoid being harmed. The fact that it's a bear I'm threatened by, and not a fellow human, would make little difference toward my feelings of danger/threat and the instinctive need of self-preservation. The antagonist's species would make little difference to my feelings of the whole attack being unjust. Sure you can reason that I was on her territory or that she was hungry, but my personal safekeeping would come first. Whether or not the bear is hungry I would still not willingly offer myself up as food nor feel like my furry sister should be physically attacking me, when I know I am of absolute zero threat to her or her potential cubs. I would equally feel empathy and concern for anyone who happened upon a bear in a forest and, regardless of what's going on in the bear's mind or tummy, would still not be happy with the bear's behavior.

Of course, at this point in the plot, you are likely thinking, "But what about the hunger of the bear?" Isn't it odd, now that the assailant is nonhuman, how we immediately start preparing a verbal defense for the offender? Like suddenly because the aggressor's shape, form, and DNA is slightly different from our own, this somehow justifies their actions? If we were talking of a human aggressor here, living out on the moors, catching the occasional backpacker, then murdering and eating them, we would be horrified. I mean, it's one thing to kill someone but completely something else to then bloodily butcher and devour their corpse!

And you still insist the bear is hungry, it doesn't know any better, it feels threatened, and that it behaves in the only way it knows how. Instinctively. Well, let us put aside the bear for a moment. The first thing needed is empathy for the victim, the hapless fellow who, innocently minding his own business, stumbles into the wrong place at the wrong time. Can we at least agree that none of us desire to take on that role? And for very valid reasons! Now instead of a bear, substitute a lion, and instead of a human, a guileless gazelle that grazes its way into the picture. Is it not too gargantuan a task to feel the gazelle is also being unjustly treated?

Is it really of consequence who is doing the victimizing and who is being victimized? Sympathy should be predominantly with the victim. Predominantly, because it is easier to feel compassion for the victim, but, sympathy and compassion are naturally for us all ultimately, regardless of our habits. To repeat, it matters little to me who the victimizer might be, what shape and form he or she may currently be manifested in. If they are inflicting suffering upon others, then something is not quite as it ultimately should be. For it seems to me that, once we have judged flesh eating to be ungodly, there is little reason to not understanding this to its full conclusion. Something perhaps only a few vegans have given thought to!

One cannot be oblivious to the dilemma of the appetite of the predator though. Lions and tigers and bears (oh my!) have to eat too. I'd just rather it wasn't me and, although I likely don't know you, I'd still rather it wasn't you too, nor any other living sentient being. So what should they eat when they clearly have such sharp claws and razor-sharp teeth that are designed for occasionally ripping flesh asunder? A vision of Eden posits that they can change; they can evolve into more peaceful creatures, capable of living in harmony with their entourage. Not only can they change, but when humans finally stop idolizing such species as model earthlings—when we stop looking up to them as examples of how we ourselves should live, eat, and behave, and begin instead to be truly civil, conducting ourselves more compassionately (thus setting ourselves as shining examples to follow)—that such a change will become inevitable. And if, for whatever reason, they are ultimately incapable of evolving toward the light side, then they will likely go the way of the dodosaurs and abandon our earthly plains.

Transcendental Experiment

In the 1990s a global political party with a spiritual agenda, who called themselves "the natural law party," declared they could lower the crime rate of a city through the group practice of transcendental meditation.[136] Several demonstrations of this were undertaken, both in the US and the UK as well as other countries, apparently with varying degrees of success.[137] I mention this solely as substantiation and evidence of how the power of positive thought can have beneficial effects on the surrounding masses without anyone necessarily being aware of the transmission of those meditative cerebrations.

I'm not suggesting we all sit down and meditate on converting the hyenas, although clearly such an action would not be detrimental on any level. What is suggested though, is for us humans, in many ways a pinnacle species on this planet, to change our own ways and to become more harmonic. Merely by doing so, other species will be influenced likewise. The reverse has been shown to also be true. Chimpanzees living in reasonably close proximity of humans have learned to hunt down prey, whereas genetically similar bonobos living away from the sphere of human influence survive without the need to hunt and kill their supper.

One Thousand Monkeys

You've probably all heard of the "thousand monkeys" story. No, it's not the one where they sit down and collectively compile the complete works of Shakespeare! It goes like this. There's a group of islands at some irrelevant place on the planet, and among other creatures on these islands lives an equally irrelevant species of monkey. Their main food source is some kind of tuber, a yam of sorts, which they dig from beneath the ground. Anyhow, one day, this one unique little monkey makes an accidental discovery; he drops his precious yam into the shallow depths of a small stream, wherein, of course, he immediately plunges his arm to retrieve it. After doing so, the little fellow has a realization: his yam is no longer gritty with soil. From this day forth, he begins deliberately submerging his yams into the stream, effectively rinsing away the grit from the surface. After a while, other monkeys begin copying their buddy, until eventually the whole population of monkeys on this particular island have followed suit and every monkey there washes their food before eating it.

Up until this point, it all seems to be within the realm of the believable, but something inexplicable then follows, something beyond the realm of the readily conceivable. The monkeys on the other surrounding islands of the archipelago also begin washing their yams, despite there being clearly no recognizable means of communication between them. It was as if the time was right. Critical mass had been reached and all the monkeys collectively learned a new skill.

Critical Mass

There is a collective conscious that extends far beyond a single species, and although each of us may only be a single monkey or one of a small select group, ideas grow, and one day critical mass is attained. The whole of humanity will move forward, and with them other species will follow, until one day, paradise on this Earth will be manifested.

Yes, I'm the eternal optimist. I predict no timeframe, but my conviction is genuine. There will come a day when we will stop idolizing predators, the lion will no longer be the king of the jungle, and eagles will no longer be seen as "majestic" birds of prey. Instead, we will see these animals for the bullies and butchers they truly are.

Of course, carnivorism is not practiced solely by so-called predator animals; every species has fallen, and every species has a need to evolve. The jungle is not safe for anyone. Even the lions have their blood sucked by tics. There are many aspects of an Eden evolution that are required before any semblance of a successful harmonious balance can be reached.

We might, for example, reason that without cats and their hunting habits, rodents could easily proliferate exponentially. This is true, which is exactly why all species need to accommodate to bring about a new balance of nature. Mice, in recognizing the cat as no longer a threat, could then cease multiplying so profusely. Of course, the re-evolving of nature may appear ludicrous, and readers may be wondering if this vision is for real, but once expressed it becomes communicable. How does the saying about revolutionary ideas go? First they will be ridiculed, then they will be violently opposed, and finally they will be accepted as elementary, normal thinking.

Try not to be overly concerned about how these changes can be accomplished, and instead concentrate on changing yourself. Focus on a world without violence and you will surely be helping to make such a beautiful dream become a reality.

Leading the Way

Risking possibly yet more ridicule, it is my belief that even when a lion takes the life of a zebra, it is at some level aware of the atrocity it is committing. The lion (read tiger/bear/crocodile/spider, etc.) knows it takes lives unwillingly, and each time it does so its soul suffers for the act. I acknowledge a balance in nature, but I see it more as a chaotic dance, the dance being the balance. Instead of it being gentle and harmonious, it is violent and full of discord.

A view from the vantage of the Eden I am expressing shows all of nature to be suffering from this discord and crying out for help. Not just the major predators but all life (the plants too). Continuously competing and trying to outperform one another, everything is part of the madness. All life wants the suffering to stop; it needs it to end. Deep down all sentience wants, indeed, yearns to stop being part of this mess. But they can't. They are perplexedly stuck in a cycle and caught in a frenzied dance, a vicious one from which they are unable to break free. They desperately need guidance and for the song to change, so the dance can smooth down with it.

I see it as our duty, our role, to help change the song. To change our ways and stop the harm. To be the role models we were always meant to be and to realize this is precisely what is meant when the bible says mankind has been given dominion over life on Earth. To be its spiritual guidance counselor and to move it harmoniously forward. This is the essence of Eden Fruitarianism.

Chapter 11
Eden

A False Eden

Ask someone to describe how they envision Eden to be, and many will likely picture what they consider to be an idyllic setting, perhaps a boat upon a peaceful lake teaming with fish waiting to be hooked. Perhaps their ever-faithful dog and plenty of rabbits to be caught. Ahh. Such peace! Seen from the point of view of the fish and rabbits, however, such a setting would be far from peaceful, idyllic, and heavenly.

Peace for Everyone

However, an earthly Eden—a hypothetical future Earth scenario—would be an unspoiled peaceful haven for all its inhabitants, regardless of what species one might happen to belong to. Every living creature will be able to walk, crawl, run, and hop around freely without fear or threat. Long grass will never harbor mischievous creatures waiting opportunistically for unsuspecting victims to sink their potentially venomous fangs into.

Fangs, claws, talons, and razor-sharp beaks designed specifically to rip living flesh asunder will no longer be part of any creature's accessories. Creatures such as the crocodile will agree to leave their current forms behind in favor of other more harmoniously compatible vessels. Should the lions and tigers choose to join us, their bodies will transform and they will adopt new peaceful ways of being.

There will be no blood-sucking vampire beings, no mosquitoes, leaches, and tics, and the temperature will have a year-round consistency. No more seasons of barren bleak winter and snow-decked landscapes. No more months of parched drought followed by monsoon rainfall and hurricanes. No more tsunamis, earthquakes, and catastrophic landslides, no more floods and blight.

If I am wrong, and there are mosquitoes, we will rejoice as they sing in our ears, content in the fact they are merely buzzing their threat-less sweet love serenades.

Goldilocks Zone and Axial Tilt

Let's get cosmical! Astronomically speaking, the Earth is well positioned to make the dream of an Eden utopia become a reality (as I believe it once may well have been). We are already nestled snugly within the planetary Goldilocks Zone,[138] the astrobiological region, neither too close nor too distant from a star that will allow for a planet to maintain livable temperatures and conditions. If humans are able to start seeing themselves as true custodians of the planet, if they can clean up their act and begin being true spiritual role models for other species, then the only thing left to make Eden a reality would be for the Earth's obliquity[139] (its Axial tilt) to straighten itself up.

Although it does vary minimally, due to the influence of the moon's gravitational pull, currently, it is tilted at an angle of roughly twenty-three degrees. Were it to be straightened up entirely, with the angle reduced to a vertical zero degree, there would be no more seasons. No more harsh winters or excessively hot summers. No more torrential rainy seasons followed by months of drought. No more annual cyclone and tornado seasons. Of course, this is all supposing the Earth's forests are reinstated and that humanity and other species have their acts together.

It's a lot to ask for, you might think, but for an eternal optimist like myself, although the scenario may seem improbable, if things can ultimately be better there, let's head there! Let's not worry about things so far ahead on the path that they bear no real consequence to the current decisions that need to be made. Trust that there will always be solutions to any problems encountered on this journey! Don't pooh-pooh the idea; we live in a universe of infinite possibilities. If you believe this is an impossible dream, don't let it deter you from still moving in that direction. If it feels right, just head in that direction.

When we are ready for Eden, it will be ready for us!

Walkabout

Indigenous people of mainland Australia had this rather beautiful tradition called "Walkabout."[140] When they became of age, they would step out of their comfort zone, leave the familiar and habitual, and walk out into the world on their individual quests for adventure and enlightenment. I believe many will once more recognize the beauty of this practice, and will choose to undertake such pilgrimages also. Of course, this does not mean they will be catching and smashing turtles or battling wits against a snake they wish to flay and devour. Neither does it mean they'll know which logs to check for witchetty grubs.[141] The trees will bear fruits of all kinds abundantly, such that wherever we choose to explore we shall always have food. There will be more varieties of fruit more wonderful than the durian, more tantalizing than the chempadeks, and more sweet and yummy than the sweetest and yummiest fruits presently on the Earth today.

As we peregrinate, we will carry fruit seeds with us, plant them in appropriate places as we go, and share them with others we encounter to care for. Even when we need to urinate and defecate, we will do so consciously, on and around small fruit tree seedlings, and communicate with them in the process.

We will all be able to wander naked wherever we choose, and with each step we take new wonders will become visible to us. Because the temperatures will be constant and the environment totally safe with neither fear nor threat, we will be able to lie down anywhere to sleep.

Left-behind loved ones will be totally at peace with our departure. People may miss one another, but they will rest assured of their safety. And what a joyous day it will be when all are reunited with all the stories we will have to share. Remember, we take to company that which we gather in solitude. To make Eden a reality, we must all step out of our comfort zones and make our lives legendary!

Continuing the Vision

Fruit will be understood. We will know that while the fruit hangs from the tree, the tree is in rapture. I believe they are like bliss receptors for the tree, ravishing unquantifiable cosmic love energy into its powerful core. Each slight brush of the breeze against the fruit is like a lover's caress, tiny orgasmic sparks that spiral inwardly. Finally, exhausted from the experience, the tree lets them go. They don't have to be forcibly cut or turbulently yanked from the tree, disturbingly terminating the tree's exhilaration. With just slight nudges, the tree will bequeath its fruit to us. Don't forget, the tree wants for you to share in the bliss! Pleasures shared are pleasures doubled!

Because of the non-existent axial tilt, even the extreme poles of the planet, barring of course any precious clean waters for our aquatic brethren, will be of constant ideal temperatures suitable for the growing of the most tropical of fruits.

The ocean's sentient life will be equally at peace. No more big fish eating little fish. No more constantly having to watch your back. Intermingling of species will be normal with no animal fearing another and no animal taking advantage of a blissful lack of fear. Because we will be in constant proximity to other life forms, the thought of capturing another and forcing it into a life of servitude through dominance and forced dependence will never enter our minds. There will be no need for harnessed pets that become reliant on us for their food and comfort. This does not mean we cannot have a special animal as a true friend, but we will never have the aberrant possessive need to cage or chain it or even to neurotically agonize over where it is at all times.

Have you ever noticed, while out walking in the countryside, how the sounds of insects and birds and nature are always off in the distance, never up close? It's like you are walking in a bubble with nature noises pretty much always beyond a roughly fixed distance. Pretty much all creatures currently fear us. As soon as we draw near, they stop their singing. As we approach they jump, run, flee, and fly away from us. If they don't, they stand stock-still in hiding. Nothing trusts us; everything goes into "danger mode" as we approach. Step toward a frog you can hear ribbitting in the foliage, and it will instantly stop.

In Eden, with the absence of such apprehensiveness, birds will be at ease singing right in front of us. Thoughts of catching or being caught will cross no one's mind. No one will be victimized for being happy and chirpy. We'll be able to walk up and look closely at the feathers of a bird and their beautiful little faces while they sing out in blissful glee of life. If they fly off, it will be because something more exciting and mind-boggling has reached their attention.

There will be streams and rivers and waterfalls to marvel at and bathe in. Survival of the fittest and this obnoxious idea of a food chain will be a thing of the bygone past, belonging to a dark age not forgotten but never celebrated.

Because we no longer fear, shun, and have this innate perverse desire to conquer nature, turning it from its vibrant greens and rainbow-colored flowers into the drab dull-grey of concrete, we will no longer suffer boredom.

In Eden, there will be no sharp, spiked plants. The roses will be thornless, the thistles too. Have you ever thought about how, after the Earth has been pillaged, its trees fallen and uprooted and topsoil laid bare, the first plants to make a resurgence are generally thorned? Take the cactus or the brambles, which readily grow into such environments. It is as if the Earth, after having being raped and desecrated, is saying, "Don't touch me there! I'm sore! Leave me alone!" And yet the Earth, as Gaia the entity, knowing our true food, still bestows us with her fruit, the blackberry and the prickly pear. Because the Earth will be healed, the thistles, spines, prickles, and thorns will be no longer be required.

Gaia currently suffers from our negligence. She is ashamed that one of her children takes what she freely gives, and greedily seeks material gain from it. And we, the supposed contemplative ones, barely think twice. We, the ones who should be planetary stewards guiding the planet into evermore color and freshness, instead trash everywhere we go. With awakened realizations, all this will change.

The flowers will be more appreciated, with their vibrant iridescence and intricate patterns, and we will all be in wonder and awe over them. Their fragrance will be ambrosial and offer sustenance for the soul.

Insects will lose their sting, for we will be more aware of their presence and less likely to brush against them accidentally. The thought of stealing their hard-earned food supplies will be unthinkable, and all compulsion to kamikazecally protect the hive will be obsolete. No animal will carry venom. No plants will be poisonous to the touch or burn upon contact.

Even the plants amongst themselves will have abandoned their competitive and invasive habits. No plant shall ever have parasitic tendencies. No more strangling and slowly sapping the life out of another plant.

Gentle touch is an entirely separate issue though. All life enjoys the sensation of loving touch. If we touch and/or are touched by a loved one, it feels good. If we stroke an animal who loves and fully trusts us, it will feel pleasure at the touch. We both will. It matters not what species of animal, once mutual trust has been gained, both will benefit from touch. It can be mammal, reptile, bird, or insect. Even spiders enjoy being touched lightly without fear of threat.

Plants too will be in bliss from attention directed toward them. In Eden we will all be at ease and comfort with touch. No one and no thing will be a threat, so the touching of spirits will lift us up as if by wings. And each mind we encounter will be a different hue. Like rainbow snowflakes. Minus the chill factor.

The topsoil will be an ideal thick, rich humus that retains moisture, making it a perfect sprouting terrain for young fruit tree saplings, and enough rain or morning dew to render the requirement for human irrigation null and void.

Even the voices of beings will change and adapt to the new prevailing harmoniousness resonating better with the rest of the environment. Think of how currently communication of certain animals is not exactly pleasant; a barking dog can really play on one's nerves, and not in a good way! If dogs wish to continue as a species, this will change; if they wish to keep their voices, chords will be realigned. The volume of voices will change too, such that every voice will blend perfectly with the euphonious symphony of nature.

Eden is a bold statement of symbiosis. Everything done—every action, every thought—will be of benefit to not only the thinker and doer but also to all surrounding parties. The current parasitic balance of nature will be no more.

If, again, you find yourself shaking your head in disbelief, thinking that without the current system of checks and balances and without the presence of the carnivores their prey will breed prolifically, think again. It is clear that the changes needed to pull this off will not be restricted solely toward carnivorous habits. Life forms currently falling victim to barbed, clawed, and sabre-fanged beasts will no longer breed epidemically. Instead of going forth and multiplying proliferously, we shall all be at peace and living splendiferously. We will come to realize that it is precisely the threat of the blood-lusting carnivores that proliferous epidemic overpopulating exists in the first place!

Of course, even if there were to be a sudden global awakening, the transition to Eden would likely not happen overnight. This will likely take time and there will first be great upheaval and turmoil, similar to that which is experienced while the body is being cleansed through detox but on a much grander scale. A microcosm/macrocosm thing. The Earth has been grossly abused and short of divine intervention, so there must be time for healing.

There is no need to get too hung up on the complexity and scale of the changes needed to fulfill such a vision, and astronomically our lives are but brief gnats' farts in time. So if you are reading this in fifty, one hundred, or two hundred years from now and things have seemingly gotten worse, hold the vision; for each one of us doing so, we bring the reality of it closer.

Any idea we might currently hold of "how boring" will be lost in the awesomeness of it all. I wonder how many people have ever walked Machu Picchu and thought, "how boring!" There are waterfalls near us and I often take visitors there. Not one has come away thinking, "That was dull!"

Once harmonic order has superseded the current state of disharmonic chaos, the necessity for change will be self-evident to all, and we will all puzzle as to why it took so long.

There will be no money. No more accounting. Not even bartering. The mentality of sharing will be universally adopted by all, and exchange will be no more. Royalty, landowners, governors, politicians, and butchers will become "once upon a time, in a galaxy far, far away" enigmas from the bizarre land of cabbages and kings. It may not be blatantly obvious to most of us, but plants experience indignation when they are swapped for money. For instance, trees feel belittled and humiliated when their fruits are stripped and sold. All this is given freely and joyously, and humans have brought shame and dishonor to the Earth by such flagrant profiteering.

Naturally there will be no wars, nor struggles for dominance and power. How could there be when greediness is put aside for empathy? When faith replaces fear? When jealousy is replaced by plenitude and trust? There will be no more boundaries and border guards preventing us from walking freely in any direction we choose.

There will be no more patriotism, no more nationalism, no more countries and passports, no more neurotic, bullying border guard officials who unquestioningly follow orders deporting, detaining, and imprisoning those who wish to practice their innate birthright to travel. The only law will be to do as you wish, provided it does not intrude on the privacy nor jeopardize the health and safety of any other living being.

Any creativity channeled into devices or technology will be done so with a full awareness of consequences, such that no life is injured or dismayed in the process, and there won't be any harmful long-term ramifications of such. With the new globally evolved consciousness, parts of the brain will be unlocked such that much of technology we think of as essential today will no longer be required. Telepathy may enable us to sing to one another in our minds and to be in touch regardless of distance.

One thing is for sure in the world as it currently exists: human creativity has evolved far beyond its spirituality, and thus we have unlocked nature's secrets to create technology that is first and foremost, used by darker powers, for destructive purposes. Humankind's spirituality needs to catch up by leaps and bounds, keeping a check on technological innovation and intentions, such that if a new secret of the atom is unveiled, its destructive weapon potential is fully disregarded.

The long and the short of it is that beings are not mean to each other in Eden, there is an ampleness of everything, and nobody will try to deceive anyone or trick anyone into deals that are not to everyone's benefit. With the complete and utter absence of struggle and strife on all levels, peace fills the void.

It is a difficult task, considering the current sad state of disrepair the world is currently lost in, but we must do our bests to behave as if such a world already exists. By doing so, we help pull it towards us. Don't make the mistake of dismissing this vision, or any dream, just because its manifestation will take time; remind yourself that time doesn't stop. This is indeed a challenge, but really, how can we expect to ever live in Eden if we cannot ourselves behave appropriately?

How do I know how things are in Eden? Well, quite frankly, I don't, of course. I think I've made it clear already that the views expressed within this book are based on faith, not indisputable scientific facts. Unlike most authors of similar books, I am not telling you that anything I say is fact; I am not talking as if I have irrefutable knowledge. That doesn't make me any less confident and convinced of my stance and viewpoint. And it shouldn't make my opinions any less valid. In fact most authors of diet-related literature kid themselves and their readers in stating that their opinion is the unquestionable truth.

I'm not a gambling man, but if I were, I'd consider it a sound wager to bet my bottom dollar that if you were to turn up at Eden carrying a fishing rod and/or shotgun—or even with the mere notion that some of its free-ranging critters were potential food items or there for you to capture, subdue, and ride; or that living trees were perfect for felling—I expect you'd be politely but firmly told to move along and take your business elsewhere. If you don't like the vision I'm sharing, then find another one, or make up your own!

Paradise Restored

For the biblically orientated, there are some quite beautiful passages affirming such a vision as an eventual outcome. Parts of Isaiah are worth pointing out once more: "The wolf will live with the lamb, the leopard will lie down with the goat, the calf and the lion and the yearling together" (Isaiah 11:6), and, "The wolf and the lamb will feed together and the lion will eat straw like the ox; but as for the serpent, its food will be dust! They won't harm or destroy on my entire holy mountain" (Isaiah 65:25).

Also, Matthew 6:25 and onward is particularly relevant. It's basically to remind us we should have more faith. We should look at the birds of the air and realize how without all the sweat and toil humans are fond of, all their needs are still met. It's true, it seems like most animal species have their place on Earth (albeit not without issues) and that all their basic needs, not counting free from threat, are met. Humans, though, really seem to be an exception to the rule. Take society away from even the most able people and they will struggle to survive. Is the reason for this because our ideal environment no longer exists on this planet? Eden is where we ultimately belong. It is the environment we are best suited for. It may have been here once already, but if it has, it's clearly been trashed.

Chapter 12
The Eden Fruitarian

At time of writing, the Eden Fruitarian is an extremely rare individual, a pioneer moving against the grain; however, there will come a day when everyone on the planet will be one.

Eden Fruitarians understand the basic principles outlined in the previous chapters and know that these concepts will inevitably one day be understood and embraced by all life. They are realistic about these changes and understand that, first and foremost, we must become the change we wish and need to see.

It's a challenge few are currently genuinely willing and ready to face up to, and even fewer have the willpower, faith, and conviction to succeed. Understandably so, as it's not as if the world we live in makes things easy. All around us we are surrounded by temptation, obstacles, negativity, and opposition. The vast bulk of us have an awful lot of emotional baggage and dependency on those temptations and obstacles. Eden Fruitarianism goes against most of what we have been taught. When aired publicly, one can expect it to be ridiculed and criticized.

Well-meaning family and friends will likely do their utmost to stop you from pursuing such a path. Perhaps, and I'm musing here, it is because deep down they fear being left behind. But it is precisely fear that the Eden Fruitarian needs to forsake.

There comes a time in everyone's life when we must stand on our own two feet and choose our own direction and path. This is of course a metaphorical path, and does not necessarily mean we have to break contact with those we love and are loved by. It does, however, mean we have to learn to stand our ground and be prepared to move against the current of social norms.

Although, as history has shown, some have made headway, changing the entire world, by any one person alone, is a gargantuan task. What is well within the realm of the feasible is for each of us to change and shape our own small worlds and environments. We must seek to make our environment as Eden-compatible as we can to facilitate our path and direction of focus. For many, this may mean being brave, leaving behind the ease of the familiar, and seeking out a new home more fitting to their needs. Likely this will require faith, but that's a good thing; for each time we rely on faith, we are bound less and less to fear.

Given that the world is as it is, certain behavioral facets of Eden Fruitarianism may not be practical immediately, but this should not stop us from focusing on other achievable aspects, nor should we fall into the trap of using those tricky details as excuses to hold us back from following other qualities more readily obtainable and doable.

Firstly, as budding Eden Fruitarians, it is our duty to bring our diets more into accord with Eden principles. This involves eating as high on the ethical food tower (outlined in the Eden Fruitarianism chapter) as we are able. We should realize that animals are neither there to satisfy our gluttony, nor to sadistically imprison for testing purposes, nor to butcher as potential clothing items, nor to imprison as visual entertainment in zoos and circuses, nor as household company, nor to be part of any other industry reliant on their captivity and abuse.

We do our best to avoid all unnecessary chemicals, whether they are on our foods or in household cleaning products. Fruit has cleansing properties both externally as well as internally. It feels great to massage mango or papaya peel into one's skin and then jump into a freshwater stream! Lemon juice can be a great substitute for shampoo, especially when one first begins to break free from the regular shampooing most of us have grown accustomed to. A daily rinse in water is all one ultimately needs to keep hair clean and the head itch-free.

Eden Fruitarians recognize the body as the temporary sacred temple of the soul. They understand it is our duty to keep it internally clean and clutter-free. Among other things, this will help us to better appreciate the resounding universal hum of OM. We have a far greater responsibility toward our body than the owner of an expensive piece of machinery needing safeguarding, regular adjustment, and high quality fuel. They understand fruit to be the highest quality fuel best suited for human physiology, and will go out of their way to ensure such fuel is always steadily, readily available. They understand that wine is not fruit juice and that all alcoholic beverages should be avoided, even in small, sporadic quantities. Equally, other mind-altering substances are shunned, in favor of the euphoria to be had from the intricate fragrance, taste, and texture of the likes of durians and other exotic fruits that are mere fractions of what will come to be.

Externally the body is ours to protect and watch over too, and this does not mean we should territorially carve our initials onto it with tattoos, nor punch holes through it with piercings. Nor does it mean we should regularly paint graffiti over it, such as in the form of makeup. The idea of plastic surgery, to somehow enhance the body's form, is also clearly seen as a ludicrous, sacrilegious concept that sadly too many of today's earthlings with apparently low self-esteem feel drawn toward.

Anyone can embrace Eden Fruitarianism at any stage of their life though, so don't despair. Just because you may have made mistakes in the past, does not mean that you cannot move forward, regardless of previous ignorance, blunders, and slip-ups. Can anyone here truthfully say that their behavior is totally compatible with Eden, and that they live according to all its terms and conditions? Many of us though, may surely sorely love to be given a chance in the genuine Eden. So, don't be overly discouraged by the difficulties, and realize the forever-giving universe is ever forgiving and that above all it is up to us to forgive ourselves and move on with optimism. We should laugh at our mistakes and see them as important lessons, and try our darnedest to not make the same ones again. Even if we keep failing!

Eden Fruitarians strive to leave cities and towns behind, and are instead naturally attracted to more scenic settings where naturism and sun worshipping can be practiced. Naturally, by "sun worshipping," I am referring to the practice of enjoying the sun's radiant energy soaking into every pore of one's body. Eden Fruitarians know fresh air, sunlight, and clean water are important foods for the soul, and enjoy all three at every available opportunity.

Instead of taming our head and facial hair growth to conform to a society so clearly in need of healing, Eden Fruitarians are happy to let their hair grow, regardless of gender, and the folly of forever fighting back the beard is regarded with bafflement.

Being aware that our food is rarely of the quality it should be, Eden Fruitarians see the importance of growing as much of one's own food as possible. They are not the slightest bit shy or afraid of getting their hands soiled in the Earth as seeds and seedlings are planted and transplanted.

Eden Fruitarians abandon all footwear in preference to baring their feet to the elements. They recognize the harmful effects of a long-term shoe fetish, and love the feeling of regaining contact with the Earth and the notion of permanent foot-zone therapy.

Eden Fruitarians are not at all concerned with calories, carbohydrates, and colostrum. They don't fret over vitamins, minerals, and proteins, or question whether or not tomatoes should be eaten because of a vague rumor that they contain nicotine. They don't call fruit "sugar" and label avocado as "fat" leading to an artificially restricted intake of either. They don't look up to someone else for advice on what, or when, or how much they should eat. They have an intrinsic understanding of the guidelines needed with concern to eating, (outlined in the Eden Fruitarianism chapter) and embrace the simplicity of it all.

Eden Fruitarians need neither cookers nor microwaves, nor a cacophony of pots, pans, and baking trays. They understand the difference between store-bought carton juice and genuine fresh juice, just as much as they get the difference between a shriveled prune and a plum. They generate very little in the way of rubbish, and most, if not all, of what passes through their hands is able to be composted or recycled in some fashion.

They understand the sometimes-overwhelming finiteness of life and that all things come to an end, and do their best to not be overly attached to material objects. They understand that we are more than just our bodies, and that there is a great white spirit, the Divine, through which we are all intricately interconnected, that all life is fundamentally one, and that harm to a part is harm to the whole.

Eden Fruitarians don't go running to the doctor at the first sign of discomfort. They have an understanding of health far exceeding that of the corner store general practitioner, and don't need someone to sell them drugs to suppress uncomfortable self-inflicted symptoms or to diagnose the obvious.

They don't smear chemical concoctions all over their skin before moving into the sunshine and don't confusedly label coconuts, sugarcane juice, and similar as fruit.

They trust that the path to freedom lies not through science but through faith. In fact, they see that we all rely on faith. They place their faith in the Divine rather than science. This ultimately gives far more peace of mind, as instead of being fundamentally in denial and under the impression that we actually know things we really don't, it requires admitting one's faith and surrendering oneself to it.

They understand that if a law is unjust, it is our moral obligation to disobey it, provided we can do so without jeopardizing our own safety or the safety or well-being of other living entities. An "unjust" law encompasses all that has no real value in the eyes of Divine.

Chapter 13
A Brief Summary So Far

My feeling is that there are plenty of books on vegetarianism and veganism already written far more eloquently than I am capable of writing. There are more books too on the benefits of eating raw and eating predominantly fruit, all of which have their merits and explain in far more detail than I am attempting to, and most of which wield whatever science there is to back up their claims. The fundamental difference between this book and others is the vision I've just shared.

I believe Eden Fruitarianism goes beyond what others have previously expressed and will likely challenge most readers in some way. I am happy for that.

I want to challenge. I want to make you question things you have never questioned before. I want the shod omnivores who read the first chapters to have "aha!" experiences and to see the need for change. I want the vegetarians to realize that vegetarianism is not enough. I want the vegans to see the sense in ceasing to fire up food before consumption. I want the vegans who recognize violence as abhorrent to question whether nature's current way should be admired, just plain ignored, or envisioned differently. If one thinks that the human race would be better off without violence toward other humans or other animals, why should we not draw the next logical conclusion and question violence within nature? If one thinks violence should be intervened and prevented within the realms of human activity, should we not also begin to debate and challenge violence amongst other species?

I want the raw food-ers and the budding fruitarians to understand fruit from more than just a dietary perspective, for everyone to see the potential it has beyond its rudimentary undervalued status as snack food. I want everyone to see the pet trade as a flagrantly depraved business without any animal's welfare as its ultimate goal. I want the shodarians to take their shoes off, to feel the Earth, and wriggle their toes! I want clothing to be put under the spotlight and questioned. I want people to open up their bathroom cabinets and question every item they see there.

All of these desires are not based on whimsical wishes alone. There are good solid reasons to support such changes. Once you begin to question all this, with an open mind, including the current paradigm of nature's balance, then the vision of Eden can become so much more than just a hopeless romantic fantasy. Even if you believe the vision will accomplish nothing, at least acknowledge that it's a beautiful one, which can do no harm!

What would be sad is if someone were to begin reading this book and think, "Hey, there's a sense to this empathy lark," but then, when they get to the envisioning, which they are not prepared to share, they shortsightedly choose to dismiss everything else based purely on that.

There will indubitably also be readers who will immediately dismiss the mere idea of nature's balance changing, deeming the notion to be preposterously impossible, and yet still they will want to voice indignation that I am threatening their world as if I have a very real chance of challenging the status quo. They will protest the thought of their winters being removed, claiming that to be their favorite time of year.

I acknowledge the enjoyment some may reap from winter sports, and can myself idolize the picture of a romantic log fire in a warm and cozy log cabin surrounded by snow-bedecked pine trees. Or a hot sauna followed by a quick roll in the fluffy white stuff. No denying the fun in that!

For the world to move forward though, especially toward Eden, we must let go of such attachments in favor of more harmonious ones. When we free our imaginations, there will be things far more awesome to witness, share, and participate in!

Reap what you can from these pages. If they can assist you to stop eating flesh, to cut out dairy, to cease cooking, go for it. Don't be deterred by views beyond those you are currently able to agree with! Every step in the right direction will help bring more harmony into the world.

Section Four
VARIOUS OTHER THOUGHTS

Before our white brothers arrived
to make us civilized men,
we didn't have any kind of prison.
Because of this, we had no delinquents
Without a prison, there can be no delinquents.
We had no locks nor keys and therefore
among us there were no thieves.
When someone was so poor that
he couldn't afford a horse, a tent or a blanket,
he would, in that case, receive it all as a gift.
We were too uncivilized to give great
importance to private property.
We didn't know any kind of money
and consequently, the value of a human being
was not determined by his wealth.
We had no written laws laid down,
no lawyers, no politicians,
therefore we were not able
to cheat and swindle one another.
We were really in bad shape
before the white men arrived
and I don't know how to explain
how we were able to manage
without these fundamental things that
(so they tell us) are so necessary for a civilized society.

- John (Fire) Lame Deer

Chapter 14
Food Supply

It is true that in the highly hypothetical, equally improbable situation where the world should suddenly wake up and demand more fruit, there would not be enough. Or not sufficient infrastructure to cope with such a sudden increase in demand. But, well, so what? This is not something we need to be concerned about.

Eden Fruitarianism is in its infancy. There are a few pioneers, but mostly the world is completely oblivious to the philosophy. It's like a flower having its roots in vegetarianism growing from the dirt and compost of a chaotic shod omnivore world. It pushed up into veganism, then raw veganism and fruitarianism, and finally Eden Fruitarianism, like a lotus flower reaching for the sky. Its bud can barely be said to be forming but will one day blossom so openly that everyone will be enraptured by its fragrance and essence.

Meanwhile, as things move around to accommodate this new concept, it is up to those of us who understand and embrace it to help manifest more supply for those yet to follow. This can be done from wherever we are, merely by making our desires and demands more public—by growing more of our own food, starting our own fruit supply, and delivering to others who are interested (and you can be sure there will be). There are all kinds of people waiting out there in the woodwork who are beginning to awaken but need help and encouragement to do so. So if you are living in an area where there is not enough supply, don't just complain about it. If you are not prepared to move somewhere else, then do something about the lack. Make it happen! Make your needs known to the universe, and they will move toward you.

Meanwhile, it is unfortunate that probably none, or very, very little at most, (yes, you read that right) of the fruit and commercially available food is fully compatible with the Eden Fruitarian philosophy. Most will have been subjected to some form of toxic chemical abuse, or other non-ethical processes, including human exploitation and genetic modification.

The most widely available alternatives of organic produce are often almost equally despicable, with heavy reliance on mammal blood and bone or crushed fish paste sprays as the basis for fertilizer. Fruit trees also, both conventional and organic, are regularly and systematically pruned causing much destruction in the process. More on this shortly. It stands to reason that all of these methods are far from ideal, and should not even be considered vegan! While these often seem to be the only available alternatives, and until we explore new forms of more harmonious agriculture, whatever we choose, our current system will in some way be a compromise.

Vegaculture

Likely we've all heard of permaculture? Though the idea itself was only christened in 1978, (derived from PERManent and agriCULTURE[143]), its principle tenet being the support of people and the environment, its basic guidelines have been used by an increasing minority throughout all cultures and eras. Take the time to read up about it and you will surely not fail to see its many advantages and truths. Since first hearing of permaculture in the 1980s, the idea has held its fascination with me. However, there is one aspect of it that has long troubled me and caused me to question its ideals. It is their reliance on the exploitation of animals.

This is where Vegaculture[144] comes to play, coined by my good friend Zalan Glen. It is permaculture with the added magical ingredient of animal empathy. VEGAn-permaCULTURE: basically growing one's food without the use of either chemicals or reliance on the excrement and carcasses of animals kept solely for exploitative purposes. Thus the guiding principle is one of support of people, animals, and the environment.

Don't jump to a false conclusion and think animals are absent, chased out, or discouraged from being in a Vegaculture garden though, far from it. The aim of Vegaculture is not only to produce ethical clean food but also to do so in an empathetic way, mirroring nature's diversity, thus encouraging and assisting local fauna in finding homes within your garden. It does not involve restricting their movements, stealing their eggs, or forcibly removing another's offspring so we may drink their milk.

The trouble with the controlled inclusion of animals within one's garden and food production lies with more than the obvious exploitation of said animals. Granted such small-scale scenarios are far more ecologically viable than modern factory farming techniques and avoid vast amounts of animal suffering, but they are still part of the problem, not the solution. Modern day factory farming has been extended from this traditional base, so in a way, permaculture is a sort of step backwards, with the question of exploitation not fully being addressed. Disregarding sentience, one might easily choose to step forward into factory farming again. Vegaculture is a true step forward. By including the welfare and freedom of animals in the design process, we can put an end to blatant disregard of animal lives.

Fruit Trees

The value of trees to the environment can never be overestimated. Many may see them solely for their potential value as timber, or for the real estate value of the land they flourish on, but they furnish so much more of what we need to survive. As well as providing air/oxygen, water, and food, trees prevent accelerated soil erosion; maintain temperature by reducing the intensity of the sunshine; and stop heat and water evaporation from the topsoil, slowing down the winds and encouraging worms.

With climate change becoming a major consideration, planting more trees is essential, as they are obviously one of nature's primary and most rapid carbon storage methods.

They may take longer to begin yielding, but fruit trees provide more food per square meter of ground than any other food on this planet, with potentially much less labor involved in the cultivation and harvesting. Not only do they produce the most food but they also require far less labor and far less soil invasion than the production of any other food. The turning of the soil carries many deaths in its wake; both animal and plant life fall victim of the process.

The insanity of the world has seeped into all spheres, and fruit production is certainly not without its troubles. Over time, we need to all get involved at some level with primary industry, and help guide and steer it away from its harmful mono crop tendencies and general heavy reliance on chemical unsustainability. We could begin by getting all streets lined with fruit trees, making fruit freely available for all.

Or better yet, follow Seattle's lead with their recently opened, community-powered Beacon Food Forest Permaculture Project,[145] a seven-acre city park filled with all kinds of fruit trees and more. The park is open to anyone and everyone to come and go as they please, and for anyone to eat freely from the garden. Such a great project! This is the first public food forest in the US. Who knows—maybe if its success continues, it will be the first of many, sparking other cities to do likewise. This kind of project needs to be implemented all around the globe.

Tree Pruning

For modern day commercial fruit growers, pruning is generally a vicious and violent affair. Deafening, lumbering, crunchy contraptions wreak havoc through the orchard, compacting the soil unduly, while their indestructible fast-flying blades annihilate anything in their path. Nests, eggs, and chicks go flying, as do any other unfortunate tree-dwelling animals. All in order to unnaturally shape the tree, convenient for later harvest. This regular process is just as proliferous in nonorganic orchards as organic ones.

Lasting, powerful, truly symbiotic relationships can never be formed with a tree when one person (or a small team) put themselves in charge of thousands of trees. Even hundreds are too many—yet another reason why more people need to return to primary industry and begin cultivating and growing their own fruit food.

I'd like to mention Masanobu Fukuoka, author of the book that partially inspired the permaculture movement.[146] One little story he mentioned within that book was how as a child he had always questioned how his father would annually prune the orange trees in their orange orchard. Masanobu couldn't understand why it was necessary, and his father explained that if one didn't do so, the tree would not fruit. Later in life, after his father passed away and he had inherited that orchard, he decided to put this theory to the test, and stopped the annual prune. Sure enough it only took a year or two to see the orange yield plummet, just as his father had correctly predicted.

Instead of pruning once more though, Masanobu decided to continue his experiment. Over the years, the branches of those trees became convoluted and tangled, and through his experimenting many of the trees died. However, still not fully understanding why so much work was involved annually, he pushed on, this time experimenting with new trees that never got pruned at all. The result was that they initially took longer to begin fruiting, but once they did, the trees were giving as much, or more fruit, than the regularly pruned ones.

Masanobu drew the conclusion that pruning was only really necessary once the tree was already tampered with (i.e., once it was already accustomed to regular pruning, or if the tree had been grafted). He supposed that pruning should not be forsaken entirely, but to be done minimally, by removing any dead branches and keeping the general shape of the tree the way nature intended.

I am wondering if the reason a tree puts out more fruit after heavy pruning each year is because at some level it feels threatened. Subconsciously fearing for its life, it puts out fruit in order to continue its heritage. I've seen many cases where trees had stopped fruiting altogether, and a human has come along and decided to bulldoze them out. In two such cases the trees' root systems were stronger than anticipated, resulting in the farmer abandoning the quest, leaving the trees half pushed over.

The results were quite astounding. The following year, both trees fruited copiously! This is perhaps similar to when humans who have abused their bodies so much through bad diet can't help but spend much of their day obsessing about sex. Because deep down they are aware they are slowly killing themselves.

Richard St. Barbe Baker

Another wonderful chap worth mentioning at this point is Richard St. Barbe Baker. He understood far better than most the important role trees play. I mention him in passing as I was listening recently to an old radio broadcast from the late 1970s with him[147] (he was around 90 years old at the time. Well worth listening to!).

Manifesto: Planting fruit trees opens our hearts to nature's wisdom. Nurturing them gives exposure to life's vulnerabilities and teaches how to build ecological and human community.

Fruit is the one food that is offered to us "free of charge." There is no taking of life, but instead the fruit is offered as part of a symbiotic pact between plant, animal, and Earth. The plant remains intact and benefits from the interaction. It gets to reproduce itself. What bliss, what joy! Not only does it get to reproduce through the sowing of its seed but also the more delicious the fruit, the more chance we will nurture the seed, feed, water, and care for it, until and after it brings to bare its own fruit.

Many people have never planted a fruit tree as they think they do not own land. But we are all children of the earth, and we share equally in the valleys, hills, rivers, and seas. Let us no longer be deceived otherwise.

Phantom fruit tree planters have no formal organization, no joining fee. The earth alone holds its membership list throughout time. To belong, simply plant a fruit tree without expectation of material gain, help care for existing ones, or hold in your heart what they mean.

Is it too long to wait for an apple tree to grow? We do not need to live so fast that only instant results satisfy. Remember, we plant fruit trees not only for ourselves but also for those who will follow. Life gives unto life. By doing so we can enjoy with clear conscience that which has been handed down from the past.

With due sensitivity, plant fruit trees on any land and in any place where they have a chance of surviving. We don't need permission. Nature sows without asking, and we are all part of nature. Reconstituting the world and working toward paradise is a duty and a right that extends beyond so-called legal concepts of ownership.

Do not worry too much about losses. Accept these as part of the process. Take heart knowing other fruit tree planters are also at work. What matters is not individual success or failure but the overall process we share in.

Chapter 15
Transitioning

The Individual

When I think of transitioning, two seemingly contradictory Chinese proverbs spring immediately to mind. The first: "Great chasms cannot be jumped by two or more medium sized leaps," and the second: "Even the journey of a thousand miles begins with a single step." It's irrefutably true, a great yawning chasm exists between shod omnivorism and Eden Fruitarianism, and probably most people who are brave enough to attempt the jump fail or spend years hanging on by their fingernails on the other side.

I've seen people blogging online, writing of their failed attempts and of their binges on pizzas, burgers, and such. Now, although I think it's great they are so open and honest about their recurring lacks of success, I see that it's pretty clear they need to go back and rethink their strategy. I think it's an incredibly big ask for anyone to leap such a wide chasm, and it's likely that only the very, very rare, most ready will ever succeed outright. For a raw food vegan, of course, the chasm has narrowed somewhat, and such a leap becomes much more feasible.

A healthy approach is to view Eden Fruitarianism not so much as a goal but as a path whose ultimate destination is the utopia of Eden—a path bound predominantly by ethics and compassion, without which one can soon go astray.

By attempting to walk the path before one has even really succeeded at enlarging their circle of compassion to embrace animals, one is doomed to fail. One has to first really—and I mean really, really—realize that eating meat is comparable to paying someone to punch your next-door neighbor in his face while you watch and enjoy. In fact, I under-exaggerate; it's far worse! There are stun guns and knives and running blood and cold steel and prodding and cold jets of water and despair and feces and urine and panic and foreboding and tortured moans and more suffering than you can shake a stick at. And that's just the last hours of an animal's life. If you're stuck there and still feel diabolical urges to eat zombiely, you need to educate yourself more, and maybe visit a slaughterhouse.

Maybe become vegetarian or vegan first, but understand that there's a history behind that pizza that you are responsible for! Once the understanding has fully sunk in and taken root, it should be impossible to look at a burger in the same way. When you know there is no going back, re-evaluate fruitarianism or consider a raw vegan diet. Find others at the same stage, online and offline, and hook up with them so you can encourage each other and laugh and cry at your mistakes. Support is a good thing. Knowing you are not alone with your struggles can be of tremendous benefit to everyone. If you are genuinely committed to Eden Fruitarianism you will work on narrowing the chasmical gap and one fine day you will succeed in bridging it.

Many self-christened fruitarians still seem to be unaware of their worldly ethical relationships. They still haven't done much with their circles of compassion; they are in it for health and fitness alone and their attitude is one of who really gives a damn if they might occasionally choose to eat a fish or a turkey.

The fish does. The turkey[148] does. This is really the predominant reason I've coined this new term, Eden Fruitarianism, to make it clear that this is not about weight loss, healing, or fitness. Although, to be sure, all three are much sought-after byproducts. Eden Fruitarianism is NOT a diet.

It is, above all, an awakening. It is not a goal but a path toward ever-increasing freedom, a path that cannot be followed without widening one's empathy and increasing one's understanding of ethics.

Addiction

A great deal of emotional investment is attached to cooked foods and foods further down the previously mentioned ethical food tower. Understanding emotional attachment is crucial to breaking free of it. Above all we should do our utmost to stay out of the bottom three layers: the flesh, the sentient life byproducts, and the grains. Although many may fall into the trap, try not to let yourself be fooled into viewing the likes of nuts, dried fruits, oils, and super foods as being valid Eden Fruitarian fare. Be aware of the instinctive stop. The subtle changes in the tastes and textures of food indicating it is time to push them aside. Once you have learned that commercial nuts, dried fruit, oils, and super foods are void of the stop, you will better understand why they should be avoided, which will thus better aid you in resisting their temptations.

The lower foods are often so devilishly addictive that we can easily mistake the hold they have over us as being the body crying out for proper sustenance. Remember it is rather like the cravings experienced by anyone coming off any heavy drug. Don't let lame and poorly thought through excuses deter you from your quest! If you find yourself irresistibly craving a food not eaten as offered by nature, then be sure that the craving stems not from any genuine bodily need but from addiction!

Understand that after defiling our internal temple, often beyond recognition of its one time splendor, there will, by necessity, be a period of detox. Many will likely also stumble at this stage and mistakenly believe detox to be a sign that the Eden diet is not working for them. This is probably the biggest deterrent of all, stopping many in their tracks and making them announce to the world that they tried but realized it just wasn't for them. It's like their house is full of junk and waste and once they start to clean it out, all manner of cockroach, vermin, and critters are disturbed from nesting amidst the piles of old pizza cartons and discarded crusty underwear. You had no idea those critters were housing in there and once they start running around everywhere, you exasperatedly put up your hands and stop cleaning. Just look at the mess you've made. If you stop cleaning altogether, the vermin will settle back down again and things will return to a seemingly relative tranquility. Yes, you are right, (I never fake a sarcasm), the dustpan and brush were a bad idea! Deep down, some will understand this, but deny it and still see it as another good excuse to quit.

Coping with the Madness

As we awaken, the knowledge of the world being steeped in negativity becomes blatantly obvious. Wars, murder, corruption, self-inflicted disease, poverty, injustice, mass slaughter of indisposed species bred purely for gluttonous purposes, babies left to cry "to teach them lessons" in goodness knows what, detrimental addictions, and more. The list continues indefinitely with the madness inescapably surrounding us. It can be easy to become overwhelmed by it all, to become affected so deeply that we lose ourselves in the depravity of it all, lowering the quality of our own lives in the process. We may experience anger, frustration, hatred, pity, and shame, or even feel so overburdened that we fall into fits of depression, and in so doing, we become lost.

Focusing on the negativity helps no one, least of all oneself. The key to success, and another great challenge for the Eden Fruitarian, is to be empathic without throwing one's own antagonistic thoughts into the equation. This will do nothing to ease or improve anything for anybody. There is no use in us being overwhelmed by empathy such that sadness rules our life. Somehow we have to keep a positive attitude despite the negativity. To see the madness and move beyond it. To acknowledge its presence and do what we can to step out of it.

The only consciousness we need to change is our own. By vanquishing mindful negativity and our own destructive habits, we can encourage and pull others forward. If we can stay clear-thinking and happy, despite the perplexities surrounding us, we set ourselves up as "examples" for the world to look up to. People will wonder, "What's his/her secret?" They will be inspired. If things get too heavy, and at times they might, you need to step aside and focus on brighter things. See the treasures. Smell a flower, hug a good friend, or go for a walk in the mountains. Close your eyes and relish the aromatic piquancy of a good durian. Lose yourself in the moment of bliss! Center yourself in the now!

The Collective

What does this movement toward Eden mean in a broader community context? More than once, I've heard people's concerns over the quandary, "What if we all turned fruitarian tomorrow; how could the world cope?" Actually, when I first heard this question it was, "What if we all turned vegetarian tomorrow; how could the world cope?" What would happen to all those cows and pigs and chickens and sheep? What would happen to all those dogs and cats if we suddenly forsook the pet industry altogether? People will argue that it is precisely because of these industries that such animals have life in the first place, so surely this is a positive thing, right?

Wrong! There is nothing "right" about any industry existing purely for profit, without genuine regard for the welfare of its sentient commodities. Do you think your life would be worth living if the entirety of it was spent incarcerated, if you were forcibly separated from your mother and family as a baby, fed poorly, had your movement restricted to the point where your muscles atrophied, where your life was ended prematurely after a long, hot, grueling, packed ride with others all awaiting the same bloody destiny? There is no point to, or yearning desire to have for, a predictable life void of freedom and any potential for inner growth. Don't try and kid yourself into thinking there is and that you are somehow doing anyone a favor by having them experience this.

It's likely that, back when slavery was being challenged, there were supporters of the practice who would have argued the same thing. You're putting slave owners out of business! Where would all the slaves go? What would they do without our support! No one can describe with any certain clarity how the world will look while the reshuffle of order takes place. Big changes are needed and the fate of livestock animals must be affected. One thing is clear already: there is a gross overpopulation of farm animals, and global statistics are not exactly easy to find, but it's estimated that the number of cattle alone is well north of a billion. In Australia there are just under 30 million[149] (The number of people is less than 23 million!) I recall that the statistics from Norway, back when I lived there, were somewhat similar, with roughly one cow per inhabitant. Five million all up. No one really keeps tabs on chickens or how many of them there are worldwide, but in Australia alone over 500 million chickens are raised and slaughtered annually.[150] The vast majority of them never even see the light of day! The approximate number of annually slaughtered pigs in Australia is five million. Anyone who continues to argue that we are somehow doing any of these beings a favor by bringing them into existence and "letting them live" under the conditions they do, has clearly never really given this any considered ethical thought.

By stepping out of the picture, and refusing to purchase consumer items necessitating bloody murder, you are not only washing your hands to it all but also helping clean up the planet and the mess it finds itself in. Worrying about hypothetical "what ifs" serves no real purpose. If the whole of human society, the collective consciousness of mankind, should suddenly wake up and see the truth and have to deal with it, then I say let such a hypothetical world deal with such an improbability. Trust that if it were hypothetically wise enough to experience a revelation of such magnitude, it would surely be wise enough to deal with its consequences too! We need also to be aware that the biggest issues will likely not be so much about a shortage in the production of fruit but more to do with the distribution of the fruit. There is likely already more than enough fruit produced to feed the world, but greed and poor organizational skills prevent it from being distributed effectively. Globally, mountains of fresh produce are composted daily!

There is a passage in the bible[151] that states that toward the end, the world will suffer like a woman in labor; this will be inevitable, but once the time has passed, life will be so joyous that all such suffering will soon fade in light of the bliss ahead. Okay, I may be taking liberties with the interpretation, but I call it how I see it and that's how it makes sense to me.

Conclusion

Many will likely consider this chapter incomplete. I know the frustration of transition and the strong urges to fall back on old habits. Having passed through it all, I understand one might want further guidance of what to eat, when to eat, and more, but the purpose of this book is to confirm our own self-reliance, and to show that such questions are for each of us alone to decide. This book identifies the path and offers inspiration. It is for you to seek further inspiration wherever and from whoever and whatever you can.

Ultimately individuals have to deal with their own issues. Misinformed choices have led us to the situation we are in. Try to find others sharing your vision, connect with them however you can, and encourage each other. Hang in there! Grin and bare it! There is light at the end of the tunnel. Don't despair; be aware! If you genuinely feel a commitment to Eden Fruitarianism, you will find your way through the obstacles. I have every confidence that success on this path is up for grabs for anyone who can truly understand the message I am sharing and who truly desires to attain the goals they might set.

Chapter 16
Topsy-Turvy World

Life on Earth is currently inherently ailing, failing, and flailing. This is something I believe we are all aware of to some extent. But because we are born into this sickness, a sickness so proliferous that it hides within itself, the true extent of such awareness is limited. Just how deep the ailing goes is mostly unrecognizable. Instead we come to consider the "ailing" as perfectly "normal" and that which differs greatly enough from that norm to be both extreme and abnormal. This is what I call the topsy-turvy world. Where reality has moved so far from the truth, that the truth has become a lie and the lie a truth, leaving us all a confused, disturbed, and basically troubled species, forever struggling with cognitive dissonance.

When I first became a vegetarian, there was a general feeling amongst my then peers that I may have been a little over-sensitive and prone to exaggerating issues of morality. When I became vegan, I had gone totally over the top. I had become an extremist. On seeing the sense in raw, and finally fruit, I most surely had lost all credibility from virtually everyone I knew. If I hadn't been of such otherwise sound mind, many would no doubt have wondered whether I was safe, loose, and caring for myself.

Of course, from my own perspective, with each new and blissful revelation, I was able to see with ever-increasing clarity just how intensely insane my original non-thought through existence had been. From the stance of the shod omnivore, Eden Fruitarianism will undoubtedly be judged as an extreme insanity by most. From an Eden Fruitarian perspective, shod omnivorism is even more insane.

What was once considered normal and sane has become abnormal and insane, and that which once would have been considered insane is now perfectly normally sane.

When we are accused of extremism, we can be quite sure it is often because the accuser has run out of better oppositional arguments. One may be considered extreme by ceasing the intake of milk and eggs, despite the well-proven evidence of this being both a sound health choice as well as an indisputably valid ethical one. Unable to logically counter these facts, opponents will thus often hide behind the torpid argument of "extremist."

The ideas expressed within this book are likely to be so distant from the reality perceived by most people that they will find it difficult to agree with any of it. I think it's sadly often the case that those who have enlightened, compassionate views and future visions are accused of borderline insanity, are ridiculed, and get criticized for thinking positively and recognize this as a symptomatic of the world gone mad.

I understand that. I face it every day. That is why I try to center myself with a non-attached humor about it all. I wish there were a way for me to share my vision, such that people could understand that although distant from the reality of today, it is not without merit, and very worth striving for. I am a realist though, despite how you may think contrarily, and understand that unless one is ready, these words will fall into deaf ears or blind eyes.

Sense of Smell

The things that we typically might think smell good become foul-smelling once one's body and mind are cleansed enough.

Naturally also (as is the very nature of Topsy-Turvyness), certain things that once smelled terrible will become more appealing than we ever thought possible. For an Eden Fruitarian to walk past a shop's decaying, cadaverous, meat and fish section, the odors are particularly, powerfully, unpleasantly pungent. However, the sweet fragrance of a durian is relished above many others. For many shod omnivores, the reverse is often quite true. Because the body is in a bad state of disrepair, all manner of sense inputs are confused and disorientated as a reflection of one's latent physiological un-wellness.

Fresh bread, probably considered by most to be healthy and a basic daily necessity, has an odor almost universally accepted as appealing. We are taught from a very young age how health-beneficial bread is, especially wholegrain dark bread, and it's not at all easy to question that loyalty. However, once we do, and have broken the addictive hold it has upon us, the attraction of the smell will gradually diminish over time until it becomes, as it rightly should, both unappealing and repulsive. Understanding the environmentally destructive nature of mono cereal crops should help to free us from any magical hold it may otherwise have on us.

There's also the confusing biblical "daily bread," and the equally bizarre biblical prophecy of "the land of milk and honey," symbols of rich fertile land for many, but taken literally, so very distant from a fruitarian Eden. The real rich fertility is to be found in the land of fruit and sunshine!

Treats

In order to incite our children to behave well, we often use the psychology of "treats."

Stop crying and you can have some cake. Be a good boy and you can have an ice cream. I've worked so hard I deserve a bar of chocolate and a nice chilled glass of champagne to swill it down. I've been so good eating raw all week. I deserve some roasted potatoes.

If only the truth were unveiled and the bigger picture more visible. We are brainwashed into thinking of such foods as something good, something we earn, something positive, so far from the addictive health-detrimental reality of the unbiased truth.

False Labeling

We have this big thing about "organic food" where it must be clearly labeled as such. This surely shows the extent of our fall. Surely it should be organic food that is "normal," and all the other foods should be labeled and marked with whatever chemicals and poisons that were diabolically added and used during their production?

Narrow-mindedness?

The extent of the madness goes far beyond what we eat though. It goes far beyond being told we are inflexible and narrow-minded, when we as Eden Fruitarians know the reverse is really true and that we would never have reached Eden Fruitarianism had we been anything remotely resembling inflexible or narrow-minded! It goes with all aspects of society that we initially regard as normal, purely because "that's the way it is."

Makeup and Body Paints

Take makeup as an example. Somehow we have been sold the lie it makes us look more attractive. This idea is so ingrained into the mass consciousness of humanity that regular daily application is considered by many to be perfectly normal. People absorb toxicity through their lips and facial skin, only to look messed up and scummy from the process. Such is the perspective of the Eden Fruitarian. Suddenly you recognize the king is in fact naked!

Now, if someone massaged papaya skin into their face, or mango skin over their body, there would be stifled, embarrassed laughter from those who will see such insanity as profanity. In such cases though, one would not even leave it on all day but follow shortly with a quick dip in fresh flowing water! Once more, the normal is abnormal and the abnormal, normal.

Nudism

Nudism, and the art of soaking up sunlight is conventionally viewed as slightly eccentric at best, to downright permissive and degenerate at worst. However, the nudist considers it a liberating and health-giving pastime and behavior to be expected on hot summer days. We also see the perversion inherent in the often-shared mass psyche that open nudity somehow automatically equates to public sex.

The Real Mafia

Then there's hair and shoes and appearances and laws. Don't get me started on laws, created by the world's greatest criminal organizations. So cleverly have they infiltrated our existence, that we've been led to believe that we can't thrive without them. I am of course talking of the governments and politicians, the most corrupt venture on the planet. Making criminals and prisoners of those who oppose. Banning us from walking freely by enforcing artificial borders. The opposite to such a state of affairs is seen as "lawlessness" and anarchy, often thought to equate to widespread uncontrolled violence and pillage. Sadly, the shod omnivores might be on to something there. We have collectively created an environment where the absence of Big Brother would be too challenging for most.

A shepherd is needed to guide his sheeple flock. Yet another topsy-turvy point, as most think of the shepherd as playing a kind beneficial role, such as the protector, the guider, the provider. And although he may be all of those things, it is only to an extent and not without ulterior motive. Cut to the chase, and the one true reason the shepherd cares for his flock is exploitative. He does not keep the wolf at bay so you can live a long, prosperous, and happy life of liberty. He does so that he alone may reap the benefits of the sheeple flesh and hides. The shepherd guiding his flock is such a poor cliché!

No, Eden Fruitarians see beyond the need to be governed, and strive to live in a truly unfettered realm. Many will rule them unrealistic and having their head in the clouds, but they acknowledge once more the topsy-turvy nature of such realism. Much as the imminence of global change (for the better) may indeed be slim, continuing to dream the way the masses do will only serve to push us closer toward the brink of annihilation. By adopting Eden Fruitarianism, one is effectively moving on and looking positively toward a brighter, more peaceful future where the only valid laws are spiritual ones—to treat all life with the respect it deserves and that we ourselves may rightly also expect the same.

But back to the government, who insists it is illegal to print your own money, to create your own passport, to sell drugs, to bear arms, to imprison, and to kill. That is of course if you are not part of, or affiliated with, that organization. It matters not what other names they may go by—the Illuminati,[152] the Bilderbergers,[153] the trilateral commission,[154] or the knights of the scimitar[155] —the crux of the matter is they are a self-appointed legalized mafia crime syndicate that have made it seem like resistance is futile and that we are powerless to oppose them.

Peddling Drugs and Vaccination

Peddling drugs is an illegal activity, and yet through their so-called "health institutions" they are peddled daily. And then there is alcohol, which is considered by most to be a perfectly acceptable and normal addition to one's social life. On the other hand, those who sell and use marijuana are considered criminals and punished, and sometimes quite severely so, for their belligerent behavior. And yet, although deaths related to alcohol intake are comparatively common,[156] similar deaths related to marijuana are virtually unheard of! Marijuana is said to be illegal because it is supposedly dangerous and health-threatening, but the real reason is because it cannot be controlled easily by the mega-corporations. If it were purely a question of health risks, the consumption of both tobacco and alcohol would also be prohibited.

Don't get me wrong; I'm not by any means trying to condone marijuana or any other mind-altering drug. I am merely showing the insanity inherent within the collective consciousness of society—the persuasion that leads to acts of legalization, to make easily available and push one of the most toxic drugs[157] onto the masses, while users of comparatively less harmful ones are victimized.

While still on the subject of drugs, yet another example of topsy-turvyness springs to mind, once more stemming from the practices of the mafia (that is, the government) supported drug peddlers. Vaccines.[158] The shod omnivorous masses tend to just accept what they are told and unquestioningly herd their children to be jabbed. With unwavering faith many will themselves queue up to receive their annual flu injections. All it takes though is a little research[159] to reveal the darker side that things aren't quite as the powers that be would have us think. Delve deeper, and you'll find vaccinations to be not even vegan friendly, with all manner of animal abuse behind their development and constitution.

Dave Mihalovic, a Naturopath specializing in vaccine research, has knocked together a downloadable form that he encourages parents to print out and present to their children's pediatricians for signing prior to having them vaccinated. It's a physician's warranty of vaccine safety form.[160] So far, he claims no doctor has been willing to do so.[161]

Sunlight

And sunlight, now there's something. We are taught, in a way, to fear it. To seek out the shade instead of the light. To go over to the dark side! As a result, reports are now coming in that health is being jeopardized through low vitamin D.[162] That's a good thing though, right? It means the sale of Vitamin D pills will increase!

TV and Media

I will use media and TV as a final example. We have become so conditioned to thinking about and accepting violence as normal that virtually everything we read and see is based around it—from the nature programs that are mostly about how animals hunt or are hunted, how they band together in flocks for safety and in packs to terrorize, how the wildebeests play Russian roulette crossing the crocodile-infested river, and how the hyenas single out the weak of the herd before ripping it to shreds. Many get their kicks out of such shows of ferocity and aggression. Fictional TV usually includes someone getting killed at some point, and often it's the focal point of the whole storyline!

Chapter 17
Compromises

The mind realization of Eden Fruitarianism is a momentous one, but after the dust of the moment has settled and the task of following its path looms dauntingly ahead, it is easy to lose oneself in the difficulty and immensity of the undertaking involved. Virtually every aspect of human existence, indeed existence in general, is currently, at least partially, exploitative and contrary to a vision of Eden. This may make the path seem futile, but this would just be lazy thinking inspired from the dark side. We need to move into the light regardless of how dull it is and how often it wavers, and understand that the more we seek it, the stronger its luminosity, and the less it will flicker, making the path ever clearer.

Meanwhile though, there will be unavoidable compromises one must make. To begin with, it is quite likely that one finds oneself in a "normal" job, in an urban setting, and often with a daily commute to and from work involved. One's choice of food may appear at first to be poor and a hindrance toward pursuing one's dream, or addiction so strong and peer/family pressure so powerfully influential that positive changes will struggle arduously against the grain of everything and everyone around you. Like you are swimming upstream and not getting anywhere.

There are two main aspects that need to be understood with compromises. The first is to not let them deter you, and the second is to not use them as excuses to compromise yet further. The more you make an effort, the easier it will be to find valid solutions.

Short of abandoning civilization entirely and moving completely off the grid, some compromises may forever be unavoidable, purely because there is no clean valid alternative. Besides, realistically, moving fully off the grid will likely also entail initial compromises, as "off the grid" there is currently very little for a budding Eden Fruitarian to eat.

Not relying on fossil fuel-driven vehicles, for example, is a practical improbability. Of course, we can lighten our carbon footprint by getting around as much as practically possible with a push-bike and a sturdy set of back panniers, but we're still reliant on our food being delivered to the shops and stalls we purchase from, and the further we are from the source of our food, the less we are likely to know its history and origins.

If there genuinely isn't enough food available though, make sure you still eat as high on the food tower as you are able, and look for ways to forever improve those choices. Until you are ready to let go more (for, at a certain stage one realizes that doing so is the only valid path), this may just mean being more aware of what's out there, or simply growing your own tomatoes in a window box.

When you are ready, committed, and enthused enough, it is best to move to an area where there is more choice, or better yet, to an area where that choice is grown locally where you have land to grow at least some of your own food.

Even after you have been on the path for a while, there will likely always be compromises of some form on another. In our own personal circumstances, although we grow some of our own food, we are still way off from being self-sufficient, and live in remote enough an area (to escape the yahooing of shod zombie omnivores and the urban madness of barking dogs and incessant lawn-mowing) that we currently can't practically manage without using a car for supplies and such.

We know it's not ideal, but we don't despair about it. We do the best we can and aim toward ever lessening the necessity of fuel consumption. If this world could straighten itself out and more people grew more fruit around us, the need for a vehicle or questionable food suppliers could be diminished yet further. We all need to work together on this.

As an aside, even this book is in some ways a compromise of my morals. I'm stuck behind a computer screen aware I am doing injury to my arms and eyes and posture in the process. I am debating with myself whether or not I should have it published on paper and, although I know once more it isn't ideal, I am strongly considering doing so. I feel the message is sufficiently important enough to merit the paper, even though part of me struggles with the regret I will feel in doing so.

It is true, we still buy some fruit from supermarkets, thus some of it is not as locally sourced as would be preferred. This world makes it difficult to follow one's ideals to the T. But the important thing with compromises is to be aware of them, and to move forward with one's spirit to be one day rid of them. Remember Eden Fruitarianism is a path, not a goal. The goal is Eden.

With the current state of worldly affairs, there is generally uncertainty of whether one's food is grown, picked, and distributed correctly, unless of course one grows it oneself. There are the issues with a whole host of chemicals used in fertilizers, pesticides, herbicides, etc. Even the certified organic food has its issues—the blood and bone and similar byproducts from the animal abuse industry, the destructive mono crop methods most frequently employed, the borderline slave labor task force, working to harvest and prune, the inefficient methods of distribution, and much more. There's even the issue of dishonesty concerning the history of food. There are foods sold as organic that are not, and foods sold as local that are not.

Even if we have land and passion to grow our own foods, the highly tuned empathy of the Eden Fruitarian will soon realize there are compromises there too. For those who took the time to watch the Myth Busters video I mentioned earlier in the book,[163] or those who are already sensitive enough to be aware of plant perception, once one sits down and works mindfully on a garden, one becomes aware of actions and their effects. There is an inevitable need to weed out grasses and other plants growing in the wrong places. Short of becoming breatharian, there is no current way around this dilemma. However, through "growing it oneself" one can be infinitely more aware of all steps involved. Weeding can be minimized by using newspaper layering and mulching, communicating with the plants as one does so. They WILL hear you!

It is our task to inform ourselves, as best we are able, and to make the best choices we can based on that information. Until the world sorts itself out, the best choice is to grow whatever we can ourselves and to not let compromises deter us on our path!

Chapter 18
Benefits Of Eden Fruitarianism

1. You'll have a clean conscience with all animal exploitation, including the bred victims of the pet trade.
2. You'll be doing your part to help restore the environment.
3. Your food choice will be the least harmful of all.
4. No cooking will be involved. Thus no need for pots and pans, etc.
5. No oils involved.
6. You'll stop the harming of your body caused by eating inferior foods.
7. You'll be helping rid yourself of karmic debt.
8. You'll save time with food preparation and clean up.
9. You'll save electricity due to the no cooking.
10. You'll find yourself generating a lot less rubbish, with even the potential to completely cut it out.
11. Fruit cleanses as well as nourishes us and will eventually bring us all to our ideal weight.
12. Fruit helps keep our body supple.
13. Fruit gives us a mental clarity and nurtures a spiritual well being.
14. Fruit reduces the amount of water needed and keeps us well hydrated.

15. Unlike many of the other foods, ripe fresh fruit is completely non-addictive and always accompanied by the instinctive stop.
16. Fruit is the best tasting of all raw foods.
17. Fruit is the most colorful and pleasing to the senses.
18. Bowel movements will be predictably regular.
19. The toilet waste of a fruitarian is far less offensive and, without treatment, is a welcome addition to the soil in one's garden. (You'll have tomatoes popping up everywhere!)
20. You'll find yourself sleeping better, with dreams being less chaotic.
21. Being barefoot, you'll be more connected with the Earth.
22. And you'll be receiving permanent foot zone therapy!
23. Your energy levels will increase.
24. You won't need all those cleansing products, or can at least avoid them as much as possible.
25. You'll have empty bathroom cabinet.
26. Suddenly 90 to 95% of the supermarket will become irrelevant.
27. There'll be virtually no regular need for doctor or dentist intervention.
28. No more annual/biannual colds. (The germs have no lasting interest in a clean body!)
29. No need for barbers and hairdressers. Perfumes, deodorants, and makeup.
30. If practiced properly and from an early enough age, it should ensure a long and happy life.
31. All pain will be lessened.
32. Flowers will smell nicer.
33. You will smell nicer.

Chapter 19
Miscellaneous Musings

You might be mistaken in thinking that, were I to rule the world, I would abolish many things, but such thoughts are far from my mind. I would not ban slaughterhouses, nor carnivorous animals, nor drugs, nor force the world to be vegan. Prohibition just sends industries underground, making them seedier and darker.

However, were I to be able to guide the world while it is still in the clutches of darker forces, I would encourage ethics as a compulsory school subject and support the global need for gaining and nurturing much-needed empathy through media and maybe financial incentive for scriptwriters to write enthralling stories, focusing less on guns and punches, and more on the light side.

Politics

Politically, while government structures still exist, I'd like to see some kind of enforced accountability for promises made that are not kept. Especially when voiced intentions turn out clearly to have been nothing more than mendacious vows. The simple commandment of "thou shalt not lie" should be honored by all those in positions of popular trust, and those who disobey this law, should be publicly humiliated at best. This needs to be rigidly enforced, and those it concerns need to be closely monitored at all times. Perhaps it would make some think twice before entering an occupation where one can so easily be corrupted. I am not entirely ignorant of the enormity of such an undertaking; who will police the police? I believe with enough forethought, there is a possibility of such a proposition becoming a reality that functions efficiently.

What I'd also like to see change in the political system is its method of election. Of course, the ultimate goal is to be rid of such systems altogether, but while they are still around, changes are crucial to moving forward. Instead of this "one man, one vote" system, which most of us with half a brain or more can see is a bit of a farce, I propose a "one man, multiple vote" election system. Let's take a scenario; let's suppose there are the following seven contenders: Conservative, Labor, Liberal, the Greens, the Socialists, the National Front, and the Loonies. Instead of just choosing one of them, as would be the normal "democratic" scenario, let's say I can now give each party a point between +5 and -5 and that only one can get each number. I am convinced in such an electoral system, but provided the results remain unfudged, the results would offer a much more honest representation of what the populace truly wants. A truly proportional representation! Naturally, only parties that receive an overall positive set of points can represent.

When I suggested this to a friend some years ago, explaining initially that a new system could be something more similar to the way the Eurovision song contest winners are elected, she guffawed and answered, "But there's a completely unpredictable new winner every year!" Yes, because they sing different songs every year! This should not be the case with political parties.

Also, I believe it would be quite fair to make the wages of politicians hover around the countries' mean wages. I guess I really mean "median wage," not the really mean minimum one! This would surely give them incentive to raise the median and weed out those who are purely after money and power!

Family

We have a chauvinistic family name system, which is pretty much always centered on the male of the family. Mr. Rolland marries Ms. Northcutt, who henceforth becomes Mrs. Rolland, and any children they may conceive take on the father's name too. Some of the more liberated women might keep their name, or in rare instances, the husband may adopt the name of the wife instead. However, as much as it may seem like the woman's name is taken on the surface, all they have done is taken on the male name from the mother's side, which is the father of the bride's name. This has been going on for as far as the history books show. My proposition is that we suspend with this clearly male-dominant practice and instead do something creative and fun. Let's say, when a couple marry, they are given free range to come up with a new name that is a partial anagram of their two family names combined. Thus in the above scenario, they will not become Mr. & Mrs. Rolland, but combining Rolland and Northcutt together, become something like Mr. & Mrs. Latchron. They should give it as much combined thought as they would for naming their children.

As is the case with most of my ever-so brilliant ideas, when I've tentatively suggested this in the past, people have laughed and said it would make tracing one's family tree a nightmare. I counter with two, I believe very valid, points: 1. In this current digitally obsessed era, it should be no problem to keep clear tabs on who's called what and when they changed their names etc. More importantly: 2. Tracing one's family tree is really a bit of a joke anyway, especially along the father's father's (etc.) line, as the chances of infidelity increase exponentially for each generation. Go back ten generations and you can be pretty darned sure that that guy was not really your great, great, great, etc. grandfather at all, regardless of what any paperwork says. The chances are incredibly slim. The further you go back, the slimmer the chances, so although there may be a theoretical likelihood that Louis XIV has descendants alive today, anyone who has traced their family back and found him to be their ancestor is, with all probability and likelihood, misinformed.

While you ponder over the significance of that, I'm going to jump to another branch of controversy.

Breatharianism

When I first began outlining this book, I had tentatively arranged to have a Breatharianism chapter following the Eden Fruitarianism one. After giving it more thought, I decided against doing so. Although I've undergone several fasts of varying lengths, I have always done so with the intent of resuming eating later, and thus never seriously attempted to live purely on air. I've been fascinated by the idea of breatharianism for the past couple of decades though, and although I have yet to personally meet anyone who I can confirm genuinely does not eat, I nevertheless still believe such a state of being is indeed possible.

However, despite this lacking of first-hand experience, I still feel like throwing in my two pence-worth, hence I've made room for it here, in this subsection, together with a hodgepodge of other mismatched notions. Perhaps later when I feel more certain or confident about my own viewpoint on this, I will revise this section and separate it into its own deserved chapter. Or delete it altogether.

Before you get overemotional and heated about how ridiculous the concept of living on love, light, and fresh air might appear to you, please take time out to consider the following:

Firstly, there are unquestionably some out there who claim to be living on air and water alone[164] (and undoubtedly Love, Light and Hugs too). As said, I cannot personally vouch for the authenticity of any of these claims, but nevertheless I strongly suspect there are a good number trying, and some truth to at least a percentage of those claims.

Secondly, if any mature person should decide to follow a path of breatharianism, then it is fully their own choosing. I see three possible outcomes to such a choice: 1. They may not succeed, and may die in the process; 2. They may not succeed, and return to more solid substances once more; and 3. They may succeed and move on to something higher. In any of the three scenarios, only the individual is at risk.

Thirdly, a far more grisly, greasy fact, is that those at the complete other extreme: the shod omnivores who eat the flesh of others daily do not only put themselves at risk. Indeed, given, let's say, a below-average lifespan of sixty years, it is quite (un)reasonable that a human flesh-eater might in that span of time devour the equivalent of something like 600 chicken carcasses, 15 cows, 40 pigs, 40 sheep, possibly thousands of fish, and countless other animals. I'm likely under-guestimating; one website offers the ballpark figure of sixteen thousand animals consumed over the average lifetime of an American meat eater.[165] To be sure, the average zombie dawn of the living dead flesh-eater puts the butchered burnt parts of hundreds of thousands of animals into their stomach. What a graveyard! R.I.P!

Think of all the misery and suffering this "choice" causes! This is only the fleshy side of things; include dairy and eggs, and who knows the true extent of the deaths and suffering of individuals that the average westernized human might be causing!

How many people have died from alcohol-related incidents, and in hospital beds daily from cancers, heart diseases, and Alzheimer's? All predominantly diet-related.

So how could such a breatharian state of being ever be attained, you might ask? Well, there are several theories I have come across that may shed some light. To begin with, biological transmutation seemed the most likely to me (see the section by that name in the Eden Fruitarianism chapter). Recently though, I have been corresponding with a guy claiming to be breatharian in South America. His theory is that a fully cleansed body ultimately needs no material sustenance. He is not the first to propose this, as Hilton Hotema[166] frequently mentioned much the same in his books, particularly "Man's Higher Consciousness."[167] It makes for fascinating reading.

Then there is the practice of "sun gazing,"[168] which many claim if practiced long and regularly enough can eventually lead not only to increased energy levels but also to the body no longer requiring food of any kind. Although it is easy to immediately dismiss the idea, as it sounds rather farfetched and the risk of eye damage quite serious, there is still a growing number of people practicing the technique claiming to receive varied benefits,[169] least of which being requiring less physical nourishment.

Personally, I do not recommend anyone to try breatharianism without first being of fully sound mind and body. Doing so otherwise will likely result in failure. Failure may be a likely result regardless of this, but if you are of sound mind, then you can at least cease the experiment should it be clear it is failing. In every case, I would never recommend anyone to attempt breatharianism unless they have moved to the top floor of the ethical food tower and have proven to themselves that they are able to live healthily and happily, for at least a few years, on solely fruit. No exceptions. That is, in order to step onto the roof of the food tower, to fully transcend food, fruit is likely the gateway.

Why would anyone wish to transcend solid sustenance entirely, when most people would probably look at the idea as being a painful deprivation? The thought of not eating all those "good" things one has become accustomed (addicted) to is probably not a pleasant one. Food has become such an issue in all our lives. We are probably all guilty of overeating, so, quite profoundly, how can we even expect to live without eating when we can't even live without overeating?

For me, the breatharian path is about letting go. It's about understanding we are more than just gross physical manifestations. It's about accepting our true potential and once we learn to transcend the newly proposed tower, all that is left is air, love, light, and material independence. Presumably this would give us the freedom to explore, to walk unfettered in any direction. To boldly go where no one has gone before. Well, perhaps nearly no one.

Military

I believe this idea was first suggested by Peace Pilgrim,[170] but it is worth including here, as it's still as valid. She suggested that instead of the military focusing on devastation and serial killing, the army could be responsible for monitoring and cleaning land pollution, the navy could do likewise with the waterways,[171] and the air force could clean up the airways pollution. What a great idea!

Prisons and Schools

Both are symptoms of a sick society. I don't believe in oppressing symptoms and wouldn't dream of doing so with these ones either, but it's worth acknowledging; neither would be around in a more healthy world. Crime is pretty much inevitable once you group so many people together that we no longer even know who our next-door neighbors are. If cities and large towns broke up and we all moved into smaller more tightly knit communities with more emphasis put on growing our own food (especially fruit trees) and having meaningful social interaction and ethics, then crime would inevitably diminish.

For me, school was an ordeal, often a nightmare, and I perceived we were all being trained to be good little society slaves or army drones; even the uniforms drove this home. Learning was a struggle, as much of what was forced upon us was really not very interesting or was presented in ways that watered our boredoms. This was not just an issue of having terrible teachers in oversized classrooms. The whole school system seemed wrong to me.

I'm not proposing children not be taught anything, just in smaller more manageable sized communities. I believe teaching would be an everyday aspect of living and all adult members would be a part of that process, teaching kids the basic necessities of life and whatever else they naturally have a flare for. I Especially believe that the most important subject of all should be empathy and ethics. Teach children to recognize other sentient life forms, including plant perception. Show them nature's art, which is expressed through flowers and how the plant objects if we rob it of its expressiveness by forcibly separating the flower from the plant.

In any case, while schools exist in the general form they do, wouldn't it make sense to model them around the most successful examples of such? Maybe learn from the Finns,[172] or learn from the Russian Shchetinin school.[173] Meanwhile though, in the absence of such innovative alternatives,[174] give serious thought to home schooling.

Countries and Borders

I have never considered myself by any means to be a patriot. I am not a nationalist. I do not consider myself to be English, nor

Norwegian, nor Polish, nor Australian, although I might lay down claim to any of those. I might carry a passport, but I resent doing so, and do so only because I know it has been my nature to travel and I am more than well aware of the neurotic bullying border guard officials who unquestioningly follow orders of no real value; who would probably deport, detain, or imprison me if I were to do otherwise. If I had more strength, I'd probably burn my passport publicly, as I have belief in the words of Gandhi that if a law is unjust it is our moral obligation to disobey it openly. My rather broad interpretation of "unjust" encompasses anything that opposes true spiritual values.

I do not believe in countries or boundaries. It is my opinion that we are, each and every one of us, aliens here. It just happens that geographically the region named England is the first country I visited, this time. But no territory belongs permanently to any one country. Boundaries cannot last as their divisions represent brute statements carved by military force. A time will come when there will be no politically imposed borders. "The Earth" will be our country, our home, and we will be forever aware that we are all just "visiting."

Humanity and all life on Earth have already suffered enough due to politically imposed borders. I have to live somewhere and am happy that of all places it is here in Australia. But I do not belong here. I do not belong to any one country. All nationalism is in some way or another a form of fascism and, as such, is a hindrance toward world peace. A danger to true progress. Nationalists are neurotic. Human history is ample proof of this.

One day, the goods of this Earth will be distributed unselfishly according to the needs of all life. But equality cannot be established by force; it will come from the heart. Take our children. Look in virtually every school throughout the world. You will see how openly "love of country" is taught. This generally sows the seed of misunderstanding and often even hatred toward other nations. This must and will change.

Littering

While still on the subject of cleaning up, I'd like to point out how perfectly natural it is to litter. In our lost true environment, if we pick an orange from a tree, there is nothing wrong with pealing it and throwing the peel anywhere. It will quickly be broken down and absorbed by the Earth becoming one with it. Fast-forward to the twenty-first century though and that banana skin is more likely to be a sweet or lolly wrapper, plastic trash,[175] or coke can. We fight our ingrained true nature each time we diligently search for an appropriate place to discard such. I am not suggesting we throw our wrappers just anywhere! I am merely pointing out how, because we have distanced ourselves so much, there is a constant uphill struggle to keep that distance. Work, money, debt, etc.

Money

Now, money's something we could talk of until the fat lady sings—the way it is so grossly unevenly distributed and the illusion of the banks and financial systems. Clearly, while it is still being used, something needs to be done to share the riches fairly. What would help is if somehow money could be made perishable, such that if a dollar doesn't change hands, it gradually depreciates in value. Yeah, I'm scratching my head on that one myself. I don't have it all clear in my head yet, but the idea doesn't seem completely infeasible. There should be a "maximum" wage, as well as a minimum one, maybe the maximum one could be just five or six times the minimum one.

Anything earned above that should either disappear completely, or be put toward some kind of fund whose goal is to help heal the world. I believe also it would be wise to stop inheritance, maybe not entirely, but certainly to an extent, such that the wealth in the hands of the rich families is slowly released. Certainly if anyone dies without an heir, leaving money in the bank, that money should not be swallowed by the bank, but contributed to the "heal the planet" fund. Yes, yes, I am aware of the complexity of all this, but this system of money is already mind bogglingly complex such that a few added rules would not be out of the question.

Taxes

Let us decide what should happen with tax money. If we wish to ensure no money is spent on arms and the military, we should be able to nominate more precisely how it is spent. It should be about "we, the people" and what we want, not what is decided for us.

Death

I believe it was Benjamin Franklin who said that the only certain things in this world are taxes and death. I'd question his judgment there, but that's not what I'm here for. However, having given my two pence-worth on taxes, I'll throw in another two pence on death. We should be allowed to take care of our own dead. I'm someone who isn't dying to get into a cemetery in the dead center of a town somewhere. I'd like for someone to bury me in my garden and plant a fruit tree on top of me. The idea that my body must by law be buried in a casket six feet under in a cemetery, or burnt otherwise to cinders is a clear money making scam. Why waste precious wood just to put someone's body in! Cardboard should suffice and will help the corpse to break down quicker. Or just throw the body in a hole, and cover with Earth and plant a tree!

Calendar

A reform of the calendar: start the year on the northern hemisphere winter solstice. This would seem logical enough. There could be thirteen months with every month consisting of twenty eight days starting on a Monday, (thus the first of every month would be Monday, and we would no longer have to remember how many days in each month) and New Year's day can be an extra day simply called "new years." Every leap year there can also be a "leap day" added to the end of each year. The month names can be rechristened, as they clearly make little sense as they now stand (Sept = 7? Oct=8? Nov=9? Dec=10?). This way calendars can be basically the same every year. Think how much paper is wasted from calendars that are printed and then discarded because they are not sold. Such a waste. This would not be the case if all years were the same. Americans could finally conform to the rest of the world and have their dates listed in the dd/mm/yy format!

Language

It seems a great idea to have an international language that is both one hundred percent logical and completely nonbiased to either gender or language groups and is simple, coherent, and elegant. A language that has both room for ambiguity but also can be completely clear with no room for confusion. The purpose of such a language could not only be a compulsory neutral language subject in all countries of the world but also to be a main focus for all language translation. It amazes me that Google and such have people working to translate to and from every combination of known language, when it would be far simpler to have language translation purely in to and out of this new proposed language, or a hub language. With such a concentrated focus, translation progress would be increased manifold. Simplifying global communication greatly.

Several languages exist already that partially fulfill the criteria of such a language, Esperanto[176] being probably the most known and used of them. However, although it definitely has potential to be fine-tuned, it fails through being European language and gender biased. More recent attempts like lojban, although quite logical, leave no room for ambiguity, which is a necessity for successful translation from a language that is fundamentally ambiguous.

For example, I wish the English language contained a neutral third person pronoun that wasn't "it." One that could be used instead of "he" and "she" to describe a living person or being where gender was unimportant

Alternate Fuel

Mechanically minded I am NOT. I've heard rumors that Tesla[177] came up with some free energy source that could be transmitted wirelessly,[178] and I'd definitely vote for that if it were vouched for as safe, authentic, and no longer suppressed.

Meanwhile though, in the absence of such enlightened technology, I believe a simple, relatively eco-friendly, solution for fuelling vehicles could be quite easily obtained from water and sunlight. I would not myself know where to start, but nevertheless I have enough knowledge to recognize this as quite readily doable. Basically it would involve solar roof panels for a free stationary power supply. This power could then work on extracting the hydrogen from water fed to it. The hydrogen could be channeled into high-pressured tanks, which then in turn could be connected to a modified lowly combustion engine and used to fuel one's vehicle. All problems you might foresee with such an undertaking could be overcome. Fossil fuels could very quickly and efficiently become a thing of the past.

It seems pretty self-evident to me that we have this massive battery of energy called the sun that likely gives more energy in a single day than could possibly be needed to keep humans going for far more than an entire lifetime. So the question needs to be asked, what have all the real geniuses out there already come up with, and what has been suppressed?

Modern-day alternatives are impending environmental catastrophes. Whether it be inevitable oil spillage[179] or the even more destructive Chernobyls and Fukushimas (said to currently be spilling three hundred metric tons of radioactive waste into the ocean daily).[180] They are Pandora's boxes just waiting to be cracked open, and with each one hope shines ever dimmer...

Positive side note: Peru is reported to be offering free solar energy to the country's poorest inhabitants.[181] Bravo, Peru!

Jobs

Over the course of the past couple of centuries, the percentage of population working in primary industry has decreased manifold. The number now is at an all-time low, with barely even 10% of the workforce involved in food production. This ever lowering of primary industry is an ever-increasing disaster. Everyone needs to re-evaluate the work line they have chosen, as most jobs in today's world are clearly worthless. Indeed, many jobs are less than worthless and outright harmful to Gaia.

This is something we all need to wake up to. This whole idea of all gainful employment being positive and blatant unemployment being negative, when clearly neither is necessarily the case, needs to be rethunk.

Conspiracies

It's my opinion that the media often doesn't report the truth, and when it does it frequently only reports it partially. We are lied to so often, so cleverly, on such a large scale, that for many the lies are blatantly obvious and those with such insight will inherently distrust all printed and reported media-filtered words (especially anything to do with politics, the military, terrorism, and health, most news of which generally have much bigger stories and ulterior motives behind them). Yet, despite the blatant lies, there is still a large percentage of the populace who blindly buy into all reported words, trusting them unquestioningly and believing the conspiracy theorists to be nothing more than overly paranoid, tinfoil hat crackpots. I believe the public is rarely ever made openly aware of the government's true agenda or of the true extent that money and power has corrupted the controlling elite.

Probably the biggest conspiracy out there concerns food, which has been transmogrified so much that it no longer bears semblance to anything that nature ever intended for us. All manner of addictive chemicals are added to hook long-term customers and make them unwilling to even consider reflecting too much on the subject. I know there are many people who as soon as the word "vegetarian" is mentioned, will express hatred toward it. The last thing they want to do is have their habits challenged! We need to learn to open our eyes, ears, and hearts, and question everything. I've already said that I'm happy for you to challenge this book. I want everything challenged!

Music

Music is another powerful tool that can be used for good or evil. Soft, gentle, notes can soothe the soul and bring more peace into the world. Contrarily, harsh, chaotic music can do the opposite, increasing more suicidal chaotic thoughts. Special concern needs to be taken to avoid "The Devil's Interval,"[182] which is a combination of notes said to be so dissonant that it makes one tense and restless, and often brings on a state of extreme negativity. That's not something any of us need!

The best music will be had in Eden, when the birds all sing in harmony with the gentle breeze, and brooks hum melodiously through our orchards, when the frogs and toads sing us their evening lullabies, and the mellifluous chorus of insects cantillates resonantly.

Epilogue

Many great ideas start out as blasphemy and it's clear to me the crux of what I'm talking of here will be seen by many as just that. Blasphemous hogwash! How dare I suggest that nature can be improved? Do I think I can do better than God? What right do I have to interfere with nature? I can already sense the accusations. I propose no interference though, and in fact it is not so much about focusing on changing the world, as changing oneself. I recognize that if one focuses on what's wrong, one can easily miss out on the ever-present joys too. I know I've tried to point out the extent to which human society is mad, but beyond the realms of this book, my main focus is on what can be in a lighter more love-filled world and on bringing my own world more in sync with such an eventuality.

I don't go around telling everyone how they should live, or tut tutting at their choices; in fact, there are many people in my social sphere who have no real idea of my life choices, as well as good friends and family who do not share all my view points but we get on nevertheless. If people ask, I'm always happy to share and answer any questions, and I try and always do so maintaining a sense of humor in the process. Hence the terribly corny jokes. I think it's important to be light hearted in the face of adversity and to not let the negativity of a demented world get the better of oneself.

I've not tried to claim the guidelines I've outlined are inflexible fact. I don't know that they are. If you think this is a copout and that I'm avoiding responsibility, then you have things twisted. As stated in the disclaimer, I am not responsible for you or your concluded thoughts.

Each of us alone has responsibility for the decisions and choices we make and the conclusions we reach. Whether or not someone else thought of them first is irrelevant. I have reached my conclusions and they have proved successful with me. I am confident about what I am saying, but at the end of the day, each must find their own confidence!

I'm not saying anyone has to live their lives any given way but am only stating how I chose to live mine and how I chose to see mine progress and, yes, how I believe the world would be better off if others followed suit.

I have also not tried to claim that by us all eating just fruit we would automatically solve all the world's problems. It should be clear that many things must change before even the fruit we eat is ideal and thus Eden-compatible. However, the potential for it to be such is clearly present, and with time, focus, and effort can become just that— Eden compatible.

Although I have no doubt some will determine I am judging others for their actions, I acknowledge it is not my place to do so. My ribbing at 801010 and Doug Graham are light hearted, and not meant to offend anyone. I acknowledge that there are well-meaning people following his path, and many very aware of the circle of compassion and the need for it to blossom. Let it be known that the judgment I pass is on actions themselves and not on he or she committing those actions. Although, in retrospect, I guess this is a slight copout, because clearly I am judging. I'm ruling humankind, especially the homersapiens, and all life on Earth to be suffering an affliction of insanity. Don't worry about my judgment though, because in and of itself it is without consequence. I suggest you settle on your own equally inconsequential conclusions.

I can't stress this enough, I know you may think of me as an incredibly critical person, but my basic philosophy is really that we should all do what we truly believe is correct, so if you think this book is all nonsense, and wholeheartedly believe shod omnivorism is what this world is all about and the path you believe wisest to remain on, then that is precisely where I believe you should remain. However, if you feel offended by my critique and judgment but are unable to state clearly why, then I suspect there may be a deeper reason. Probably cognitive dissonance. I mean, why should you care what I think? If you're confident with your beliefs why should you feel threatened or offended by mine? If anything, feel pity for me. I don't care, because I'm confident with my beliefs.

If there are questions that I've forgotten to tackle and issues I've neglected to address, and I'm sure there are, try and take a step backwards and answer them yourselves. If you're wondering where honey lies in my food tower and you truly grok what I'm saying, then surely you can find a place for it. I'm hoping that Eden Fruitarianism is not just another concept that gets embraced by people who can't think clearly and who end up dragging it through the mud.

From the profound and gloomy depths of shod omnivorism, it might seem like Eden Fruitarianism is purely an emotional choice. However, the higher one eats on the proposed food tower, the more blatantly clear it becomes just how much it is truly a logical choice. So if you are reading this book a hundred years from now and nothing appears to have improved, you can rest assured the spirit of what I write is still very much alive. It is immortal and has been around since time immemorial; I have tasted from its source, and it is still as fresh and pure as ever, forever flowing, infinitely waiting for us all to join in.

About Me

Most people believe they know things, but I believe my self-professed wisdom comes from knowing I believe things. I see a profound difference between the two outlooks. I concede that my knowledge is limited and recognize I don't have all the answers. None of us do; none of us are perfect, least of all myself (and I speak from profound personal experience). Recognizing such failings though should not prevent us from moving forward. Our ignorance and inability to see the road ahead with perfect vision and clarity should never prevent us from trying to improve ourselves.

Let me very briefly introduce myself. Born in 1961, I call myself Mango, and together with my life partner, Květa, I live a peaceful life in the tropics of FNQ Australia. You might think it's all well and good that I write of these things from my comfortable life of relative abundance, but for most people things aren't so straightforward, with far too many difficulties and obstacles impeding one's progress and journey. This would be quite an unfair assessment though. I too started out life as an unwitting, zombie flesh-eating shod omnivore, in a large town (Wellingborough) in the grey, damp drabness of Northamptonshire, England. As a young teenager I began plotting my escape, and in 1982, the day before my twenty-first birthday, I fled England's shores. I've since lived in other large towns and cities, in countries with winters longer than summers, I've faced many daunting challenges throughout my journey, and impatiently yearned for the life I have now, evoking divine powers of manifestation to bring it slowly toward me. I've been impatient and frustrated when things haven't always been smooth, and the journey has not been short either, but has taken many years, indeed, decades. But each time I have experienced a revelation, I have never lost sight of its significance, and strived to ever live life more in fulfillment of such revelations. One must take on the attitude that abundance is more something we tune into rather than something that requires blood, sweat, tears, and the almighty dollar.

I have started work on my autobiography, which explains more about me and my life and revelations, my hardships and joys, and narrow escapes throughout that journey. It is barely begun, but I hope that one day it will be and that it too can offer still further insight into the ancient path of Eden Fruitarianism.

I will announce it on my blog and Facebook page[183] once it is finally available. (It could take years. This one did!) Meanwhile, here's a little self-introduction I penned some years ago.[183]

The earth is our mother. Whatever befalls the earth,
befalls the sons and daughters of the earth.
This we know. All things are connected
like the blood which unites one family.
All things are connected.
We did not weave the web of life, we are merely
strands in it.
Whatever we do to the web we do to ourselves.

- Chief Seattle

End Notes

1   "Pure Fruit," Vimeo Video, 1:01:44, documentary about Mango & Kvĕta, by Emile Bokaer, 2012, https://vimeo.com/36666602.

2   "Quite Possibly the most Eye Opening Six Minutes Ever on Film," Video, 06:15, August 16, 2013, https://www.minds.com/blog/view/201538/quite-possibly-the-most-eye-opening-six-minutes-ever-on-film.

3   Dr. Seuss, "The Lorax," Amazon.com, Accessed June 24, 2015, http://www.amazon.com/gp/offer-listing/0394823370/ref=as_li_tl?ie=UTF8&camp=211189&creative=373493&creativeASIN=0394823370&link_code=am3&tag=veganfruitari-20&linkId=3523ATUI5OWZETXB.

4   Catharine R Gale, "Research into IQ in childhood and vegetarianism in adulthood: 1970 British cohort study," the BMJ, 28 October, 2006, http://www.bmj.com/content/334/7587/245.

5   Virginia Messina, ""Empathy, intelligence and the vegetarian brain," examiner.com, June 1, 2010, http://www.examiner.com/article/empathy-intelligence-and-the-vegetarian-brain.

6   Marc Bekoff, "Brain Scans Show Vegetarians and Vegans Are More Empathic than Omnivores," Psychology Today, July 12, 2012, , https://www.psychologytoday.com/blog/evolved-primate/201005/empathy-is-what-really-sets-vegetarians-apart-least-neurologically-speak.

7   "Cognitive Dissonance," Wikipedia, Accessed June 24, 2015, https://en.wikipedia.org/wiki/Cognitive_dissonance.

8   "Abattoir worker slits puppy's throat because he couldn't afford vet's bill," Daily Mail Australia, December 9, 2011, http://www.dailymail.co.uk/news/article-2071775/How-Abattoir-worker-slit-puppys-throat-afford-vets-bill.html#comments.

9   "Brazilian Pork Sausage Piglet Prank," YouTube video, November 29, 2012, https://www.youtube.com/watch?v=NG4WhppBNCM.

10   Unknown Author, "Vegetarians live longer and have better health," Eating Ecologically, Accessed June 24, 2012, http://www.eateco.org/Medical/Vegetarian.htm.

11   Ashley Capps, "Catching Up With Science: Burying the 'Humans Need Meat' Argument," Free From Harm, July 17, 2013, http://freefromharm.org/health-nutrition/catching-up-with- science-burying-the-humans-need-meat-argument/.

12   Helen Stockebrand, "Comparative Anatomy Chart," photobucket, Accessed June 25, 2015, http://s39.photobucket.com/user/orange-light/media/debates/comparison.jpg.html.
Michael Bluejay, "Humans are naturally plant-eaters," Vegetarian Guide, Updated October 2014, http://michaelbluejay.com/veg/natural.html.
Wishro, "Science Verifies that Humans are Frugivores," Scribd, July 5, 2009, http://www.scribd.com/doc/17111888/Science-Verifies-That-Humans-Are-Frugivores.

13   George Dvorsky, "Prominent scientists sign declaration that animals have conscious awareness, just like us," io9, August 23, 2012, http://io9.com/5937356/prominent-scientists-sign-declaration-that-animals-have-conscious-awareness-just-like-us.

14   Alastair Bland, "From Pets to Plates: Why More People are eating Guinea Pigs, The Salt, April 2, 2013, http://www.npr.org/sections/thesalt/2013/03/12/174105739/from-pets-to-plates-why-more-people-are-eating-guinea-pigs.

15   Stevo, "Video Facts of meat production," meatvideo.com, Accessed June 25, 2015, http://www.meatvideo.com/.

16   Gary Youofsky, "Best Speech You Will Ever Hear," YouTube video, December 22, 2010, https://www.youtube.com/watch?v=es6U00LMmC4.

17   Sheeplecorporation, "The 'Earthlings' video, YouTube video, April 17, 2011, https://www.youtube.com/watch?v=19eBAfUFK3E.

18   Keith Akers, "Animal Brothers, 4th Letter," compassionatespirit.com, Accessed June 25, 2015, http://www.compassionatespirit.com/animal_brothers_4.htm.

19   Mark Hawthorne, "The Mental Health Consequences of Killing For a Living," Occupyforanimals, March/April 2011, http://occupyforanimals.wix.com/meat#!slaughterhouse-workers/czcf.

20   Chris Winter, "The Case for Vegetarianism," YouTube video, June 4, 2013, https://www.youtube.com/watch?v=sJNntUXyWvw.
Conufan, "Adorable Irish Girl Explains Why She doesn't want to eat meat," YouTube video, July 5, 2015, https://www.youtube.com/watch?v=mMgJYQgYS24.

21   The slave industry is still very much in existence. In fact, there are said to be more people enslaved now than ever before in history. The difference between then and now, however, is that the current slave trade is vastly invisible and all takes place without the consent of the current global legal systems.

22   Free The Slaves, "Slavery is Everywhere," freetheslaves.net, Accessed June 26, 2015, http://www.freetheslaves.net/about-slavery/slavery-today/.

23   Doris Lin, "How Many Animals are Killed Each Year?," animalrights.about.com, Accessed June 26, 2015, http://animalrights.about.com/od/animalrights101/tp/How-Many-Animals-Are-Killed.htm.

24   "Slaughterhouses," Wikipedia, Last Modified June 16, 2015, https://en.wikipedia.org/wiki/Slaughterhouse.

25   Noam Mohr, "Animal Death Count in US," animaldeathcount, Accessed June 26, 2015, http://animaldeathcount.webnode.com/all-animals-2011/.

26   ADAPTT, "Real time Kill counter," ADAPTT, Accessed June 26, 2015, http://www.adaptt.org/killcounter.html.

27   Isha Datar, "Why your burger should be grown in a lab," CNN ireport, August 9, 2013, http://edition.cnn.com/2013/08/08/opinion/datar-lab-burger/index.html?hpt=hp_c4.
Laura June, "Your Meat Addiction is Destroying the Planet," The Verge, August 13, 2013, http://www.theverge.com/2013/8/13/4605528/your-meat-addiction-is-destroying-the-planet-but-we-can-fix-it.

28   Kaita Moskvitch, "Modern Meadow aims to print raw meat using bioprinter," BBC News, January 21, 2013, http://www.bbc.co.uk/news/technology-20972018.
Ryan Whitman, "3D Printed meat could soon be cheap and tasty enough to win you over," GEEK News, February 12, 2013, http://www.geek.com/news/3d-printed-meat-could-soon-be-cheap-and-tasty-enough-to-win-you-over-1539410/.

29   Edward Immel, "Evolution of the words vegetarian and vegan," Gentle World, July 28, 2012, http://gentleworld.org/whats-in-a-word/.

30   Irish Vegetarian Society, "Vegetarian and Vegan Definitions," Vegetarian Society of Ireland, January 25, 2014, http://www.vegetarian.ie/definitions/.

31   Sara Talyor, "The Benefits of extended breastfeeding," The Daily Mom, May 16, 2016 2013, http://dailymom.com/nurture/the-surprising-benefits-of-extended-breastfeeding/.

32   Animals Australia, "Dairy Cows Fact Sheet," Ainmals Australia the voice for animals, Accessed June 26, 2015, http://www.animalsaustralia.org/factsheets/dairy_cows.php.

33   VM Sathish, "AL Ain 'Super Cow' gives100 litres of milk a day," Emirates247, June 21, 2011, http://www.emirates247.com/al-ain-super-cow-gives-100-litres-of-milk-a-day-2011-06-21-1.403953.

34   Voiceless, "Australia Battery Hen facts," Voiceless – The Animal Protection Institute, Accessed June 26, 2015, https://www.voiceless.org.au/the-issues/battery-hens.

35   Unknown, "Facts about caged egg-laying hens," nocagedeggs.com, Accessed June 26, 2015, http://nocagedeggs.com/caged-eggs.htm.

36   Amy Remeikis, "'Free range eggs' definition scrambled, The Brisbane Times, July 29, 2013, http://www.brisbanetimes.com.au/queensland/free-range-eggs-definition-scrambled-20130728-2qshq.html.

37   Animals Australia, "The Truth about cows, dairy and leather," Animals Australia Unleashed, Accessed June 26, 2015, http://www.unleashed.org.au/animals/cows.php.

38   Vegan Mainstream, "A Fur Farm Experience," vegan Mainstream, Accessed June 26, 2015, http://www.veganmainstream.com/2011/11/30/vegan-means-being-a-voice-for-animals-a-fur-farm-experience/.

39   The Vegetarian Resource Group, " Why don't vegans use leather, silk or wool?," The Vegetarian Resource Group, Accessed June 26, 2015, http://www.vrg.org/teen/leather_silk_wool.php.

40   Emily Moran Barwick, "Is Wool Vegan? Is it humane?," Bite Size Vegan, Accessed 26 June, 2015, http://www.bitesizevegan.com/ethics-and-morality/is-wool-vegan-is-it-humane/.

41   Vegan Peace, "Wool Facts," Vegan Peace, Accessed June 26, 2015, http://www.veganpeace.com/animal_cruelty/wool.htm.

42   Christine Wells, "Why Vegans Don't Use –Silk," Gentle World, December 1, 2012, http://gentleworld.org/do-vegans-use-silk/.

43   Mercy for our Sentients, "I am scared and don't want to die," YouTube video, May 29, 2009, https://www.youtube.com/watch?v=LUkHkyy4uqw.

44   Left by the side of the hive, the bees will feed from the bucket.

45   "Marital Rape," Wikipedia, Accessed 26, June 2015, https://en.wikipedia.org/wiki/Marital_rape.

46   "The Simpsons," Wikipedia, Accessed June 26, 2015, https://en.wikipedia.org/wiki/The_Simpsons.

47   Jong Gao, "Boys' urine-soaked eggs listed as local speciality, intangible cultural heritage," The Ministry of Tofu, March 11, 2011, http://www.ministryoftofu.com/2011/03/boy-urine-soaked-eggs-listed-as-local-specialty-intangible-cultural-heritage/.

48   Chirs Pirillo, "Chocolate Covered Grasshoppers," YouTube video, 5:12, Uploaded June 30, 2009, https://www.youtube.com/watch?v=_qJ7TLwPbh0&feature=youtu.be.

49   "The Grossest food You're Eating Every Day," WorldTruth.TV, Accessed June 27, 2015, http://worldtruth.tv/grossest-food-youre-eating-every-day/.

50   "Statistics for Chronic Disease," Australian Institute of Health and Welfare, Accessed June 27, 2015, http://www.aihw.gov.au/statistics-for-chronic-disease/.

51   Fast Food – Exposing the Truth," Cancer Council NSW, February 22, 2013, http://www.cancercouncil.com.au/wp-content/uploads/2013/02/Fast-Food-Exposing-the-Truth-22-February-2013.pdf.

52 "Chemotherapy Doesn't Work 97% of the Time - Dr. Glidden," GreenMed TV, April 7, 2013, http://tv.greenmedinfo.com/chemotherapy-doesnt-work-97-of-the-time-video/.

53 Christine Luisa, "Exposing the fraud and mythology of conventional cancer treatments," October 12, 2011, http://www.naturalnews.com/033847_chemotherapy_cancer_treatments.html.

54 Arjun Waila, "Man With Stage 3 Colon Cancer Refuses Chemotherapy and Cures Himself With Vegan Diet," Collective Evolution, July 21, 2013, http://www.collective-evolution.com/2013/07/21/man-with-stage-3-colon-cancer-refuses-chemotherapy-cures-himself-with-vegan-diet/.

55 Some years back, I coined the term "juice feasting" - I liked it because it was just that little "e" (f(e)asting) that made the difference and since then, I've noticed that in more recent years the term has definitely been gaining in popularity.

56 Herbert M. Shelton, "The Hygienic System Vol III - Fasting and Sunbathing," 3rd Revised Edition, 1950, http://www.soilandhealth.org/02/0201hyglibcat/020127shelton.III/020127.toc.htm.

57 Morris Krok, "Fruit, the food and medicine for man," Amazon, Accessed June 27, 2015.

58 Edmond Szkely, "The Essene Science of Fasting and the Art of Sobriety: Guide to Regeneration in Health and Disease," Accessed June 27, 2015, http://www.amazon.com/gp/offer-listing/0895640112/ref=as_li_tf_tl?ie=UTF8&camp=1789&creative=9325&creativeASIN=0895640112&linkCode=am2&tag=veganfruitari-20.

59 "Douceur et Harmonie: Domain Mamn Terre," Douceur et Harmonie, Accessed June 25, 2015, https://douceurharmonie.wordpress.com/. "Douceur et Harmonie," Ecovillage Networking, Accessed June 25, 2015, http://sites.ecovillage.org/douceur-et-harmonie-domain-maman-t.

60 Arnold Ehret, "Mucusless Diet Healing System" (see pages 16,-18), Amazon, Accessed June 25, 2015, http://www.amazon.com/gp/offer-listing/1884772005/ref=as_li_tf_tl?ie=UTF8&camp=1789&creative=9325&creativeASIN=1884772005&linkCode=am2&tag=veganfruitari-20. Arnold Ehret, "Rational Fasting," Amazon, Accessed June 27, 2015, http://www.amazon.com/gp/offer-listing/1884772013/ref=as_li_tf_tl?ie=UTF8&camp=1789&creative=9325&creativeASIN=1884772013&linkCode=am2&tag=veganfruitari-20.

61 "Vegan Diet impacts California Prison," Vegetarian Spotlight, January 9, 2011, http://vegetarianspotlight.com/2011/vegan-diet-impacts-california-prison/. "Saved By A Vegan Prison," Vegetarian Spotlight, December 20, 2012, http://vegetarianspotlight.com/2012/saved-by-a-vegan-prison/.

62   Emanuel Barling Jr, Esq and Ashley F. Brooks, RN, BSN, "Medical Doctors Have Little or No Knowledge of Nutrition," howtoeliminatepain.com, June 9, 2011, http://howtoeliminatepain.com/medical-doctors-have-little-or-no-knowledge-of-nutrition-because-medical-schools-offer-little-or-no-education-on-nutrition-vitamins-or-minerals/.

63   Kelly M Adams, "Status of nutrition education in medical schools," NCBI, June 18, 2008, http://www.ncbi.nlm.nih.gov/pmc/articles/PMC2430660/?tool=pubmed.

64   Ben Goldacre, "What Doctors don't know about the drugs they prescribe," Ted Talks, June 2012, http://www.ted.com/talks/ben_goldacre_what_doctors_don_t_know_about_the_drugs_they_prescribe?utm_source=feedburner&utm_medium=feed&utm_campaign=Feed:+TEDTalks_video+%28TEDTalks+Main+%28SD%29+-+Site%29.

65   Sarah C. Corriher, "Death Rates Drop When Doctors Go on Strike," The Health Wyze Report, Accessed June 27, 2015, http://healthwyze.org/index.php/component/content/article/502-death-rates-drop-when-doctors-go-on-strike.html.

66   Ibid.

67   Emile Bokaer, "Pure Fruit a film by Emile Bokaer," Vimeo Video, 2012, https://vimeo.com/36666602.

68   "Anopsology," Wikipedia, Accessed June 27, 2015, https://en.wikipedia.org/wiki/Anopsology.

69   "Guy-Claude Burger," Wikipedia, Accessed June 29, 2015, https://fr.wikipedia.org/wiki/Guy-Claude_Burger.

70   Loren Cordain, Ph.D, "The Paleo Diet," The Paleo Diet, Accessed June 29, 2015, http://thepaleodiet.com/.

71   Mark Sisson, "Primal Blueprint 101," Mark's Daily Apple, Accessed June 29, 2015, http://www.marksdailyapple.com/primal-blueprint-101/.

72   Dr. Douglas Graham, "Home of The 80/10/10 Diet and Dr. Douglas N Graham," FoodNSport.com, Accessed June 29, 2015, http://foodnsport.com/index.php.

73   Graham, "The 80/10/10 Diet," Amazon, Accessed June 29, 2015, http://www.amazon.com/gp/offer-listing/1893831248/ref=as_li_qf_sp_asin_tl?ie=UTF8&camp=1789&creative=9325&creativeASIN=1893831248&linkCode=am2&tag=veganfruitari-20.

74   "Seed dispersal methods," Wikipedia, Accessed June 29, 2015, https://en.wikipedia.org/wiki/Seed_dispersal.

75   Scott Zona, "Plants that use Ballistic seed dispersal," Fairchild Tropical Botanic Garden, Accessed June 29. 2015, http://www.virtualherbarium.org/gardenviews/GoingBallistic.html.

76   Christopher Bird & Peter Tompkins, "The Secret Life of Plants," Amazon, Accessed June 29, 2015, http://www.amazon.com/gp/offer-listing/0060915870/ref=as_li_tf_tl?ie=UTF8&camp=1789&creative=9325&creativeASIN=0060915870&linkCode=am2&tag=veganfruitari-20.

77   "Mythbusters – Plants have feelings(primary perception)," YouTube video, Published October 16, 2012, https://www.youtube.com/watch?v=fStmk7e9lJo.

78   "MythBusters," Wikipedia, Accessed June 29, 2015, https://en.wikipedia.org/wiki/MythBusters.

79   "Secret Life of Plants and Telepathy," YouTube video, Uploaded August 22, 2010, https://www.youtube.com/watch?v=MPpURDTTf8k.

80   "The Secret Life of Plants –Full," YouTube video, Published on October 9, 2012, https://www.youtube.com/watch?v=sGl4btrsiHk.

81   Taffine Laylin, "New Study Shows Plants Talk to Each Other Through the Soil," inhabitat.com, May 23, 2014, http://inhabitat.com/plants-talk-to-each-other-through-a-messenger-in-the-soil/.

82   R. Jenness, "The Composition of Human Milk," NCBI, July, 1979, http://www.ncbi.nlm.nih.gov/pubmed/392766.

83   "World Class Bodybuilder is RAW VEGAN," YouTube video, Published July 3, 2012, https://www.youtube.com/watch?v=2dM1iU5e33k.

84   "Vegans Bodybuilders dominate Mainstream Competition," Vegan Bodybuilding & Fitness, July 27, 2013, http://www.veganbodybuilding.com/?page=article_robert_austin_2013.

85   "The China Study," Accessed June 29, 2015, https://en.wikipedia.org/wiki/The_China_Study_%28book%29.

86   Ross Horne, "The Health Revolution," Amazon, Accessed June 29, 2015, http://www.amazon.com/gp/offer-listing/0959442383/ref=as_li_tf_tl?ie=UTF8&camp=1789&creative=9325&creativeASIN=0959442383&linkCode=am2&tag=veganfruitari-20.

87   Sophie Egan, "Making the Case for Eating Fruit," The New York Times, July 31, 2013, http://well.blogs.nytimes.com/2013/07/31/making-the-case-for-eating-fruit/?_r=2.

88   Mango Wodzak, "The Dangers of Unripe Fruit," FruitNut.Net, August 2002, http://www.fruitnut.net/HTML/205_NonFiction_Fruit_Mango.htm.

89   Derek Henry, "How to heal cavities naturally," Live Free Naturally, November 2013, http://livefreelivenatural.com/heal-cavities-naturally/.

90   "The Ayurvedic Doshas," Wikipedia, Accessed June 29, 2015, https://en.wikipedia.org/wiki/Dosha.

91   Peter J . D'Adamo, "Eat Right For Your Type," Amazon, Accessed June 29, 2015, http://www.amazon.com/gp/offer-listing/039914255X/ref=as_li_tf_tl ?ie=UTF8&camp=1789&creative=9325&creativeASIN=039914255X&link Code=am2&tag=veganfruitari-20.

92   "Food Pyramid (Nutrition), Wikipedia, Accessed June 29, 2015, https:// en.wikipedia.org/wiki/Food_pyramid_%28nutrition%29.

93   "Water facts and figures," IFAD, http://www.ifad.org/english/water/key. htm.

94   "Livestock's Long Shadow," Wikipedia, Accessed June 29, 2015, https:// en.wikipedia.org/wiki/Livestock%27s_Long_Shadow.

95   Kate Good, "How Animal Agriculture Is Draining the World of Biodiversity," One Green Planet, September 29, 2014, http://www. onegreenplanet.org/animalsandnature/how-animal-agriculture-is-draining-the-world-of-biodiversity/.

96   Tamar Haspel, "Monocrops: They're a problem, but farmers aren't the ones who can solve it," The Washington Post May 9, 2014, http://www. washingtonpost.com/lifestyle/food/monocrops-theyre-a-problem-but-farmers-arent-the-ones-who-can-solve-it/2014/05/09/8bfc186e-d6f8-11e3-8a78-8fe50322a72c_story.html.

97   Michael Bloch, "Leather and Envionmental Issues," Green Living Tips, August 24, 2012, http://www.greenlivingtips.com/articles/leather-and-the-environment.html.

98   OneGreenPlanet.org, "Facts on Animal Farming and the Environment," One Green Planet, November 21, 2012, http://www.onegreenplanet.org/ animalsandnature/facts-on-animal-farming-and-the-environment/.

99   "Eating up the World: The Environmental Consequences of Human Food Choices," Vegetarian Network Victoria, Accessed June 29, 2015, https://www.voiceless.org.au/sites/default/files/Final%20VNV%20 Environmental%20Brochure%20for%20web.pdf.

100   "Eating up the World: The Environmental Consequences of Human Food Choices," Vegetarian Network Victoria, Accessed June 29, 2015, https://www.voiceless.org.au/sites/default/files/Final%20VNV%20 Environmental%20Brochure%20for%20web.pdf.

101   "The Commonwealth Scientific and Industrial Research Organisation," CSIRO, Accessed June 29, 2015, http://www.csiro.au/.

102   "Deep-Sea Fish STOCKS," Eur-Lex Access to European Law, Accessed Jun 29, 2015.

103   "Environmental Destruction – Aquaculture Problems," International Coastal Marine Resources Laboratory, Accessed Jun 29, 2015.

104   "ISPM 15 - International Standards For Phytosanitary Measures No. 15," Wikipedia, Accessed Jun 29, 2013, https://en.wikipedia.org/wiki/ISPM_15.

105   "Barefoot," Wikipedia, Accessed June 29, 2015, https://en.wikipedia.org/wiki/Barefoot.

106   Mark Pagel, Walter Bodmer, "A Naked Ape Would Have Fewer Parasites," The Royal Society, August 7, 2003, http://rspb.royalsocietypublishing.org/content/270/Suppl_1/S117.

107   Marilyn D. Story, "Personal and Professional Perspectives on Social Nudism," JSTOR, Accessed June 29, 2015, http://www.jstor.org/stable/3812606.

108   Mark Storey, "Children, Social Nudity And Scholarly Study," o2binsxm.com, Accessed June 29, 2015, http://o2binsxm.com/childrenandsocialnudity.htm.

109   "A comparison of pro- and anti-nudity college students on acceptance of self and of culturally diverse others," NCBI, September 2008, http://www.ncbi.nlm.nih.gov/pubmed/18686157.

110   Mike Adams, "Sunscreen Causes Cancer," Natural News, May 29, 2008, http://www.naturalnews.com/023317.html.

111   "Chaetophobia," Wikipedia, Accessed June 30, 2015, https://en.wikipedia.org/wiki/Chaetophobia.

112   "Pogonophobia," The Free Dictionary, Accessed June 30, 2015, http://www.thefreedictionary.com/pogonophobia.

113   "The Truth About Hair and Why Native Americans Would Keep Their Hair Long," Science of the Spirit, September 08, 2011, http://www.sott.net/article/234783-The-Truth-About-Hair-and-Why-Indians-Would-Keep-Their-Hair-Long.

114   Mark, "45 Uses for Lemons," ALTDrudge.com, August 17, 2013, http://altdrudge.com/45-uses-for-lemons-that-will-knock-your-socks-off/.

115   "Here's What it Looks Like When you Don't Wash off your Makeup For a Month," The Huffington Post, September 17, 2013, http://www.huffingtonpost.com/2013/09/16/makeup-premature-aging-_n_3936828.html?utm_hp_ref=mostpopular.

116   Visala Kantamneni, "These 8 Countries Have Banned Wild Animals in Circuses," One Green Planet, March 19, 2014, http://www.onegreenplanet.org/animalsandnature/10-countries-that-have-banned-wild-animals-in-circuses/.

117   Taffine Laylin, "Costa Rica is Closing its Zoos and Freeing All Captive Animals," Inhabitat, August 9, 2013, http://inhabitat.com/costa-rica-is-closing-its-zoos-and-freeing-all-captive-animals/.

118   Laura Beck, "Vegan Dogs," The Bark, November 16, 2010, http://thebark.com/content/vegan-dogs.

119   Chuck Raasch, "Cats kill up to 3.7 billion birds annually," USA Today, January 30, 2013, http://www.usatoday.com/story/news/nation/2013/01/29/cats-wild-birds-mammals-study/1873871/.

120   A mouthpiece, typically made of metal, that is attached to a bridle and used to control a horse.

121   "Paro Therapeutic Robots," PARO, Accessed Jun 30, 2015, http://www.parorobots.com/.

122   Jackson Landers, "Does your pets food contain dead pets?," Slate, April 19, 2013, http://www.slate.com/articles/health_and_science/science/2013/04/what_is_in_pet_food_zoo_animals_sick_livestock_dogs_and_cats_from_shelters.single.html.

123   Nicole Adams, "Radoslaw Czerkawski to be held on bail – Read the Comments," examiner.com, December 19,2013, http://www.examiner.com/article/radoslaw-czerkawski-to-be-held-on-no-bail-puppy-doe-will-see-justice.

124   Dave Munger, "Up to Half of All Humans Are Infected by a Cat-Borne Parasite," Seed Magazine, Accessed June 30, 2015, http://seedmagazine.com/content/article/toxic_house_cats/.

125   "Worldwide Pet Overpopulation Epidemic," Dog Breed Info Center, Accessed June 30, 2015, http://www.dogbreedinfo.com/articles/breedersvsrescues.htm.

126   "Animal Overpopulation," Oxford-Lafayette Humane Society, Accessed June 30, 2015, http://www.oxfordpets.com/index.php?option=com_content&view=article&id=61.

127   Otto Spijkers, "Understanding Bokito the Gorilla," The Netherlands School of Human Rights Research, May 21, 2007, http://invisiblecollege.weblog.leidenuniv.nl/2007/05/21/understanding-bokito-the-gorilla-that-es/.

128   "Brumby," Wikipedia, Accessed June 30, 2015, https://en.wikipedia.org/wiki/Brumby.

129   "The Golden Rule," Wikipedia, Accessed June 30, 2015, https://en.wikipedia.org/wiki/Golden_Rule.

130   Daniel Burke, "The World's Fastest Growing Religion is...," CNN, April 4, 2015, http://edition.cnn.com/2015/04/02/living/pew-study-religion/.

131   Isaiah 11:6 The wolf will live with the lamb, the leopard will lie down with the goat, the calf and the lion and the yearling together; and a little child will lead them.

132   "The Lioness and the Oryx," fruitnut.net, April 2002, http://www.fruitnut.net/HTML/210_NonFiction_Animal_Oryx.htm.

133   "The Story of Little Tyke," fruitnut.net, 2002, http://www.fruitnut.net/HTML/210_NonFiction_Animal_LittleTyke.htm.

134 "Tikkun Olam," Wikipedia, Accessed June 30, 2015, https://en.wikipedia.org/wiki/Tikkun_olam.

135 Abby Miller, "Good Deeds, Not Belief in Christ, Required for Salvation Says New Pope," Addicting Info, May 23, 2013, http://www.addictinginfo.org/2013/05/23/good-deeds-not-belief-in-christ-required-for-salvation-says-new-pope/.

136 "Natural Law Party," Wikipedia, Accessed June 30, 2015, https://en.wikipedia.org/wiki/Natural_Law_Party.

137 Steven Bancarz, "Studies Reveal Group Meditation Can Lower Crime Rates, Suicide Rates, and Car Accidents in Surrounding Cities," Spirit Science and Metaphysics, Accessed June 30, 2015, http://www.spiritscienceandmetaphysics.com/studies-reveal-group-meditation-can-lower-crime-rates/.

138 "Goldilocks Zone," Rational Wiki, Accessed June 30, 2015, http://rationalwiki.org/wiki/Goldilocks_Zone.

139 "Axial Tilt," Wikipedia, Accessed June 30, 2015, https://en.wikipedia.org/wiki/Axial_tilt.

140 "Walkabout," Wikipedia, Accessed June 30, 2015, https://en.wikipedia.org/wiki/Walkabout.

141 "Witchetty Grub" Wikipedia, Accessed June 30, 2015, https://en.wikipedia.org/wiki/Witchetty_grub.

142 Matali Matali, "Manifesting Paradise", Amazon, Accessed August 14, 2015, http://www.amazon.com/gp/offer-listing/1452509751/ref=as_li_tf_tl?ie=UTF8&camp=1789&creative=9325&creativeASIN=1452509751&linkCode=am2&tag=veganfruitari-20.

143 "Permaculture," Wikipedia, Accessed June 30, 2015, https://en.wikipedia.org/wiki/Permaculture.

144 "Vegan Organic Gardening," Wikipedia, Accessed June 30, 2015, https://en.wikipedia.org/wiki/Vegan_organic_gardening.

145 Clare Leschin-Hoar, "It's Not a Fairytale: Seattle to Build Nation's First Food Forest", TakePart, February 21, 2012, http://www.takepart.com/article/2012/02/21/its-not-fairytale-seattle-build-nations-first-food-forest.

146 Masanobu Fukuoka, "The One-Straw Revolution," Amazon, Accessed July 1, 2015, http://www.amazon.com/gp/offer-listing/1590173139/ref=as_li_tf_tl?ie=UTF8&camp=1789&creative=9325&creativeASIN=1590173139&linkCode=am2&tag=veganfruitari-20.

147 Ann Jones, "Men of the Trees - Richard St. Barbe Baker interview," ABC – RN, August 26, 2012, http://www.abc.net.au/radionational/programs/offtrack/men-of-the-trees/4188212.

148 Ashley Capps, "12 Reasons You May Never Want to Eat Turkey Again," Free from Harm, November 7, 2013, http://freefromharm.org/animal-cruelty-investigation/12-reasons-you-may-never-want-eat-turkey-again/.

149 "Fast Fact 2013: Australia's Beef Industry," Meat and Livestock Australia, 2013, http://www.cattlecouncil.com.au/assets/Beef%20Fast%20Facts%202013_EMAIL.PDF.

150 "Broiler Chicken Fact Sheet," Animals Australia, the Voice for Animals, http://www.animalsaustralia.org/factsheets/broiler_chickens.php.

151 John 16:21, Romans 8:22.

152 Richard Stone, "Who are the Illuminati," Red Ice Creations, Accessed July 1, 2015, http://www.redicecreations.com/specialreports/whoilluminati.html.

153 Stephen Lendman, "The true Story of the Bilderberger Group, and What they May be Planning," Global Research, June 1, 2009, http://www.globalresearch.ca/the-true-story-of-the-bilderberg-group-and-what-they-may-be-planning-now/13808.

154 "Trilateral Commission: World Shadow Government," Illuminato News, Accessed July 1, 2015, http://www.illuminati-news.com/trilateral-commission.htm.

155 I just made that one up.. actually, it's a reference to an episode from the sit-com "Cheers."

156 "Alcohol and Public Health: Alcohol-Related Disease Impact (ARDI), Centers for Disease Control and Prevention, Accessed July 1, 2015, http://nccd.cdc.gov/DPH_ARDI/default/default.aspx.

157 Robert Gable, "The Toxicity of Recreational Drugs," American Scientist, Accessed July 1, 2015, http://www.americanscientist.org/issues/pub/the-toxicity-of-recreational-drugs/99999.

158 "Merck drug company vaccines admits injecting cancer viruses," Live Leak, November 17, 2007, http://www.liveleak.com/view?i=327_1195303011.

159 "Vaccines did not Save us – 2 centuries of Official Statistics," Child Health Safety, Accessed July 1, 2015, https://childhealthsafety.wordpress.com/graphs/.

160 Andrew Baker, "The Vaccine Hoax is Over. Documents from UK reveal 30 years of cover-up," May 10, 2013.

161 Dave Mihalovic, "You want to vaccinate my child? No problem, just sign this form", Prevent Disease, August 9, 2013, http://preventdisease.com/news/13/080913_You-Want-To-Vaccinate-My-Child-No-Problem-Just-Sign-This-Form.shtml. "Physician's Warranty of Vaccine Safety", Prevent Disease, Accessed August 15, 2015, http://preventdisease.com/pdf/Warranty-of-Vaccine-Safety-English.pdf.

162   Lorena Nessi, "Scared of the Sun – The Global Pandemic of Vitamin D Deficiency," Brain Blogger, June 16, 2013, http://brainblogger.com/2013/06/16/scared-of-the-sun-the-global-pandemic-of-vitamin-d-deficiency/.

163   "MythBusters - Plants have Feelings, Primary Perception," YouTube video, Published October 16, 2012, https://www.youtube.com/watch?v=fStmk7e9lJo.

164   Dean Nelson, "Man claims to have had no food or drink for 70 years," The Telegraph, April 28, 2010.

165   "How Many Animals Die to Feed Americans?," Animal Death Count, 2011, http://animaldeathcount.webnode.com/all-animals-2011/.

166   "Hilton Hotema," Wikipedia, Accessed July 1, 2015, https://en.wikipedia.org/wiki/Hilton_Hotema.

167   Hilton Hotema, "Man's Higher Consciousness," Amazon, Accessed July 1, 2015, http://www.amazon.com/gp/offer-listing/1169830145/ref=as_li_tf_tl?ie=UTF8&camp=1789&creative=9325&creativeASIN=1169830145&linkCode=am2&tag=veganfruitari-20.

168   "Sungazing," Wikipedia, Accessed July 1, 2015, https://en.wikipedia.org/wiki/Sungazing.

169   Vinny Pinto, "Sungazing - Observations and notes from a veteran Sungazer," Sungazing, September 23, 2007, http://sungazing.vpinf.com/.

170   "Peace Pilgrim," Wikipedia, Accessed July 1, 2015, https://en.wikipedia.org/wiki/Peace_Pilgrim.
       "Peace Pilgrim," Peace Pilgrim - Official Website, Accessed Jul1 2015, http://www.peacepilgrim.org/.

171   Boyan Slat, "How the Oceans can clean themselves," YouTube video, Accessed July 2, 2015, https://www.youtube.com/watch?v=ROW9F-c0kIQ&feature=youtube_gdata_player.

172   Olga Khazan, "The secret to Finland's Success with Schools," The Atlantic, July 11, 2013, http://www.theatlantic.com/international/archive/2013/07/the-secret-to-finlands-success-with-schools-moms-kids-and-everything/277699/.

173   "Mikhail Petrovich Shchetinin - Kin's School - Lycee School at Tekos," David Icke, March 28, 2009, http://www.davidicke.com/forum/showthread.php?t=59310.

174   Liesl Den, "11-year old shares her thoughts about homeschooling. Parents.com, March 4, 2014, http://www.parents.com/blogs/homeschool-den/2014/03/04/homeschool-den/11-year-old-shares-her-thoughts-about-homeschooling/.

175   "This Film Should be Seen by Everyone," Why Don't you try this?, November 21, 2013, http://www.whydontyoutrythis.com/2013/11/this-film-should-be-seen-by-the-entire-world.html.

176   "Esperanto," Wikipedia, Accessed July 2, 2015, https://en.wikipedia.org/wiki/Esperanto.

177   "Nikola Tesla," Wikipedia, Accessed July 2, 2015, https://en.wikipedia.org/wiki/Nikola_Tesla.

178   "Discover How to Use a 100 Year old Device to Generate Free Electricity" Nikola Tesla Secret, Accessed July 2, 2015, http://www.nikolateslasecret.com/.
"Nikola Tesla," Free Energy, Accessed July 2, 2015, http://free-energy.ws/nikola-tesla/.

179   Laura Moss, "The 13 largest oil spills in history," Mother Nature Network, July 16, 2010, http://www.mnn.com/earth-matters/wilderness-resources/stories/the-13-largest-oil-spills-in-history.
"Montara Oil Spill," Wikipedia, Accessed July 2, 2015, https://en.wikipedia.org/wiki/Montara_oil_spill.

180   Mike Adams, "Fukushima leaking 300 tonnes of radioactive water daily," Natural News, August 13, 2013, http://www.naturalnews.com/041610_Fukushima_radioactive_leak_state_of_emergency.html.
Gary Stamper, "Your days of eating pacific fish are over," Collapsing into Consciousness, Accessed July 2, 2015, http://www.collapsingintoconsciousness.com/at-the-very-least-your-days-of-eating-pacific-ocean-fish-are-over/.

181   Timon Singh, "Peru to Provide Free Solar Power to its 2 Million Poorest Citizens," Inhabitat, July 23, 2013, http://inhabitat.com/peru-solar-power-program-aims-to-give-electricity-to-the-countrys-2-million-poorest-citizens/.

182   Finlo Rohrer, "The Devil's Music," BBC News, April 28, 2006, http://news.bbc.co.uk/2/hi/uk_news/magazine/4952646.stm.

183   Mango Wodzak, "Mango in a Nutshell," fruitnut.net, May 2007, http://www.fruitnut.net/HTML/151_AboutMe.htm.